DANGEROUS DIMENSIONS

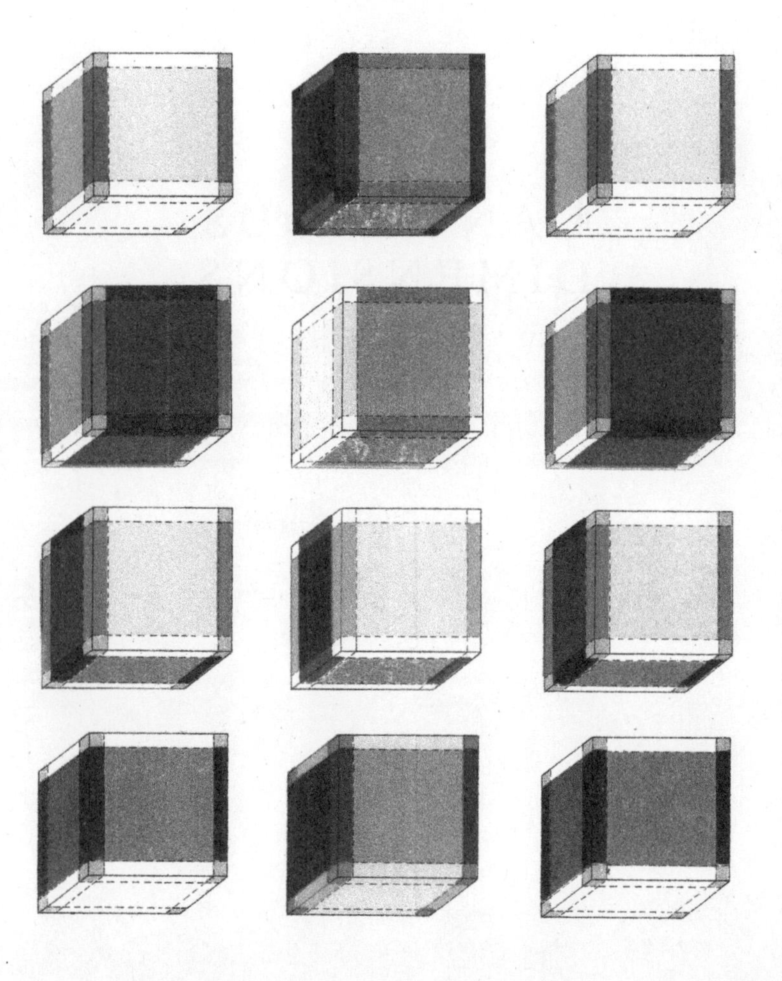

DANGEROUS DIMENSIONS

Mind-bending Tales of the Mathematical Weird

edited by

HENRY BARTHOLOMEW

This collection first published in 2021 by
The British Library
96 Euston Road
London NW1 2DB

These stories are presented as they were originally published and as such contain language in common usage at the time of their writing. Neither the publisher nor the editor endorses any language which would now be considered xenophobic or offensive.

Cataloguing in Publication Data

A catalogue record for this publication is available from the British Library

ISBN 978 0 7123 5368 7
e-ISBN 978 0 7123 6737 0

The frontispiece features 'Views of the Tesseract' from the foldout illustration in Charles Hinton's *The Fourth Dimension*, Swan Sonnenschein & Co, Ltd, London, 1904.

Cover design by Mauricio Villamayor with illustration by Sandra Gómez

Text design and typesetting by Tetragon, London
Printed in England by CPI Group (UK) Ltd, Croydon, CR0 4YY

CONTENTS

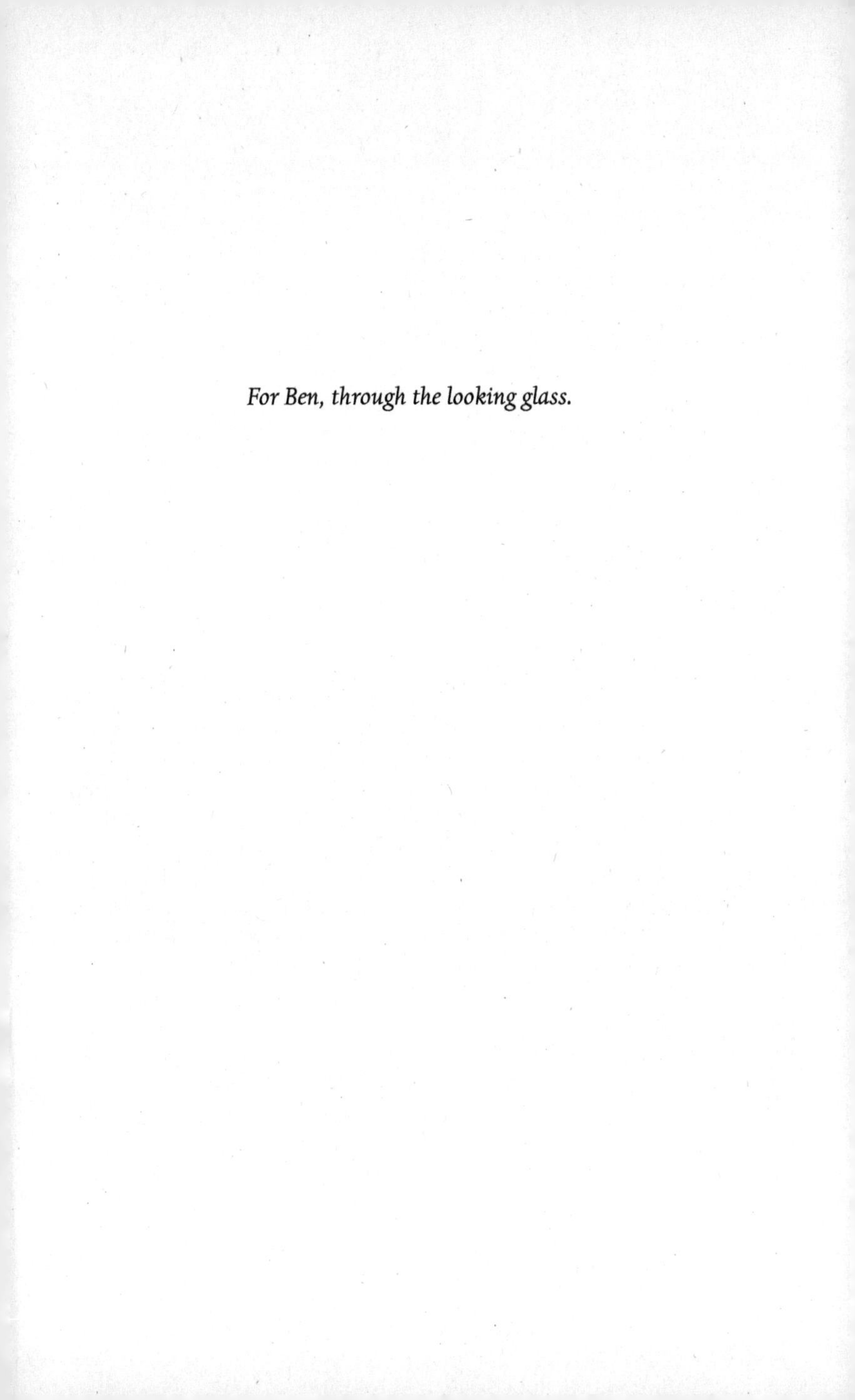

For Ben, through the looking glass.

INTRODUCTION

Unlike nineteenth-century Gothic fiction, which tends to fixate on the past, the haunted, and the ghostly, early weird fiction probes the very boundaries of reality—the laws and limits of time, space, and matter. It is well known that fictions of the morbid and the macabre have long been inspired by contemporaneous advances in science. Mary Shelley's *Frankenstein* (1818), for example, drew on Luigi Galvani's electrical experiments on dead flesh and viscera, while H. G. Wells stitched the debates and discourses surrounding vivisection into his nightmarish *The Island of Doctor Moreau* (1896). However, while weird fiction often exploits the horrors associated with biology, anatomy, and the plasticity of the human (and non-human) body, it also owes a debt to a very different endeavour: mathematics.

The stories assembled here explore the original trajectory of this influence. The collection begins in the late nineteenth century with a formative story by H. G. Wells, and ends with Miriam Allen deFord's "Slips Take Over", first published in 1964. Between these two tales are a rich variety of mind- and matter-bending narratives from a range of authors writing in the first half of the twentieth century. Whether established tale or pulp oddity, each story engages with some aspect of mathematics, geometry, or physics—always in the service of the weird and the imaginative. Some, like "The Hall Bedroom", only hint at strange mathematical formulae, while others, like "The Pikestaffe Case" and "The Living Equation", revel in the glories and dangers made possible by "higher" mathematics.

Mirrors are a recurring feature. Lewis Carroll's 1871 sequel to his wildly popular children's story *Alice's Adventures in Wonderland* (1865) famously took Alice "through the looking glass", and it is notable that Carroll, the pen name of Charles Lutwidge Dodgson, was an accomplished mathematician in his own right. Mirror-worlds, of course, are rarely benign. Long a source of uncanny power, the mirrors here are also thresholds where three-dimensional reality angles off into sinister unknown realms.

Then there's the figure of the mathematician. While women mathematicians appear throughout the history of the discipline, often making ground-breaking contributions against great odds, fictional mathematicians in the late nineteenth century and early twentieth are almost entirely men.* And while late Victorian mathematics might make one think of stuffy professors scratching innocuously at chalkboards, it was feared that living entirely in a world of numbers, logic, proofs, equations, formulae and calculations might have certain deleterious effects on one's humanity. Sherlock Holmes's nemesis, for example, the criminal mastermind Moriarty, is a professor of mathematics, and uses his "phenomenal mathematical faculty" to effect his villainous machinations.† Likewise, the nefarious intrigant "Minister D—" in Edgar Allan Poe's Dupin story "The Purloined Letter" (1844) is a poet-mathematician. In more overtly Gothic fiction, mathematicians also proved useful

* A list of these women must include, at the very least, Émilie du Châtelet (1706–1749), Sophie Germain (1776–1831), Mary Somerville (1780-1872), Ada Lovelace (1815–1852), and Emmy Noether (1882–1935).

† Doyle, Sir Arthur Conan. *The Complete Sherlock Holmes, Volume 1*. New York: Barnes and Noble, 2003. 559.

victims. Mathematicians or their proxies feature in several ghost stories of the period including Bram Stoker's "The Judge's House" (1891) and M. R. James's "'Oh, Whistle, and I'll Come to You, My Lad'" (1904), where they symbolize the arch-rationalist whose world view is shaken to the core by the intrusion of the supernatural. As the nineteenth century passed into the twentieth, however, this image began to change. With the advent of Einsteinian physics, mathematics went cosmic, and mathematicians—both real and fictional—morphed into explorers and discoverers.

This is, in many ways, a book about discoveries. As the title implies, the majority of the stories collected here concern extra dimensions. This is because extra spatial dimensions—beyond the usual three—were a topic of profound interest at the turn of the century. Indeed, the great Oscar Wilde, more often associated with *fin de siècle* aestheticism than mathematics, has his apparition, the fretting Sir Simon, "hastily adopt the Fourth Dimension of space as a means of escape" in his playful story "The Canterville Ghost" (1887).* Rummaging in Wilde's notebooks, the literary scholar Jarlath Killeen has unearthed a further reflection: "in modern science", Wilde writes, "the fourth dimension of space, infinity, eternity, &c are poetical conceptions". † As the stories collected in this volume attest, Wilde was not the only writer to sense a curious entanglement between mathematics and the imagination. That Wilde and

* Wilde, Oscar. *The Complete Works of Oscar Wilde, Vol 8: The Short Fiction*. Ed. Ian Small. Oxford: OUP, 2017. 86.

† Qtd. in Killeen, Jarlath. "Oscar Wilde in the Fourth Dimension: Ghosts, Geometry, and the Victorian Crisis of Meaning". *The Routledge Handbook to the Ghost Story*. Ed. Scott Brewster and Luke Thurston. New York: Routledge, 2018. 54–55.

other authors should be interested in mathematics at all should not surprise us. The nineteenth century was awash with mathematical revolutions and by the end of the century mathematicians of every stripe were rethinking some of the discipline's most fundamental assumptions—and opening up new worlds in the process.

The most far-reaching of these revolutions occurred in the field of geometry, where Euclid's axioms—the basis of all Western geometry for over two thousand years—were overhauled by various new systems of "non-Euclidean", that is, hyperbolic or elliptic, geometry. A string of mathematicians including Sophie Germain (1776–1831), Carl Friedrich Gauss (1777–1855), Nikolai Lobachevsky (1792–1856), János Bolyai (1802–1860), Bernhard Riemann (1826–1866), and, in the UK, William Kingdon Clifford (1845–1879)—who pre-empted Einstein's theories on the curvature of space by decades—all, in one way or another, explored the possibilities and applications of curvature and non-Euclidean geometry. The details of these new geometries, however, along with the non-orientable topologies of August Möbius (1790–1868) (the "Möbius strip") and, later, Christian Felix Klein (1849–1925) (the "Klein bottle"), would not filter into weird fiction until later in the twentieth century. In the meantime, as Wilde's notebook reminds us, it was the imaginative, poetical potential of the fourth dimension that captivated the cultural psyche rather than its precise mathematical formulation. And while the history of higher dimensional space begins in eighteenth-century France with works by Jean le Rond d'Alembert (1717–1783) and Joseph-Louis Lagrange (1736–1813), it was not until the nineteenth century and another slew of mathematicians including William Hamilton (1805–1865), Hermann Grassmann (1809–1877), Ludwig Schläfli (1814–1895), Arthur Cayley

(1821–1895), and Victor Schlegel (1843–1905), that the fourth dimension truly came into its own.

Not only could mathematics now account for a fourth dimension, it had also developed the rubric to talk about a fifth, sixth, or, indeed, any number of higher dimensional spaces: so called *n*-dimensional geometry. And this brings us squarely, as it were, to the 1880s—the decade the fourth dimension spilled from the pages of mathematical treatises to pour, tumultuously, into the popular imagination. This quickening had unlikely origins. It begins in 1884, when a London priest and schoolmaster by the name of Edwin Abbott Abbott published a story that would put dimensional geometry in the limelight. *Flatland: A Romance of Many Dimensions* tells of a square—the story's narrator—who lives in a two-dimensional world in which one's status is determined by one's shape. Sandwiched between worker triangles below him and the priest class of circles above, the narrator's steady life is interrupted by the arrival of a three-dimensional sphere. Unable even to imagine such an entity, the narrator's fellow Flatlanders silence his ravings by promptly escorting him to jail. Part geometry primer, part metaphorical commentary on class inequality, *Flatland* resolutely brought higher dimensions into the realm of Romance, which is to say, the realm of fiction.

Incredibly, and despite the originality of Abbott's story, *Flatland* was not even the most influential mathematical Romance published in 1884. That same year also saw the publication of Charles Howard Hinton's *Scientific Romances*. A colourful British mathematician, Hinton published several works on fourth-dimensional geometry in the late nineteenth century. It was Hinton who coined the term "tessaract" (now "tesseract") to describe a four-dimensional cube—a word and concept still used today in both mathematics and popular culture. By presenting his ideas in the form of stories, Hinton's works reached a far wider audience than other mathematical treatises of the time, and the influence of the *Scientific Romances* was widespread. Books like *Another World; or, the Fourth Dimension* (1888) by Alfred Schofield, M. D., claimed that Hinton's mathematics proved the truth of Christianity by explaining, among other things, the omnividence of God, angelic visitation, and even the eventual resurrection of humanity on Judgment Day. If this sounds mystical, it's because it is. While the late nineteenth century was an exciting time for mathematics, it was a veritable golden age for occultism and pseudo-science. Already in 1880, seven years before Wilde's ghostly Sir Simon was using the fourth dimension as a means of hasty dematerialization, an English translation of the renowned astrophysicist Johann Zöllner's book *Transcendental Physics* had caused a stir by invoking the fourth dimension to account for Zöllner's experiences at the séance table. This was the era of Spiritualism, Theosophy, the Hermetic Order of the Golden Dawn, and the Society for Psychical Research. The idea that there was a fourth spatial dimension, and that one might be able to travel through it—physically or otherwise—or, indeed, that the spirits of the dead might reside there, perhaps even reach out into our own world, was a potent one for occultists and authors alike.

Indeed, Theosophy—an occult society founded by the enigmatic Helena Petrovna Blavatsky in 1875—incorporated the language and speculative mathematics of Hinton's books to explain a host of esoteric phenomena, from hauntings and vampirism, to astral projection and clairaudience. Charles Webster Leadbeater, a high-ranking member of the order, went so far as to suggest that if one used Hinton's thought-experiments for adapting the human mind to higher dimensional geometry as a training regimen, one could, in theory, teach oneself clairvoyance.* A number of authors of strange fiction, including Algernon Blackwood, Arthur Machen, Arthur Conan Doyle, and Dion Fortune, were either Spiritualists or members of esoteric societies like the Golden Dawn, and aspects of these societies' doctrines and practices reappear throughout their writings. The reader will notice that the present volume features two stories by Algernon Blackwood. This is partly due to personal bias. Blackwood's fiction is unique, and the majority of his works are long overdue a reprint. But the inclusion of these stories is also a reflection of Blackwood's preoccupation with higher dimensions. Certainly, no other author of the period returned to the topic so frequently.†

Blackwood also brings us firmly into the twentieth century. Hinton's *n*-dimensional geometry, which captured the imagination of so many, was still Euclidean, still "flat". Despite its various formulations in the nineteenth century by figures like Lobachevsky

* Leadbeater, C.W. *The Other Side of Death, Scientifically Examined and Carefully Described*. Chicago: Theosophical Book Concern, 1903. 467-8. Perhaps Hinton's most intriguing legacy in the twentieth century was his influence on the thought and writings of an author included in this collection—Jorge Luis Borges, whose work also blends the mathematical and the fantastic. See Borges, Jorge. *The Total Library*. Ed. Eliot Weinberger. London: Penguin, 2001. 221, 249, 500, 508-510.

† An early story—"Entrance and Exit" (1909)—as well as two stories from his final major short story volume *Shocks* (1935)—"Elsewhere and Otherwise" and "The Man Who Lived Backwards"—might easily have been included here, but for reasons of space and parity were ultimately rejected.

and Reimann, the popularization of non-Euclidean geometry in the twentieth century is largely a result of the meteoric rise of Albert Einstein (1879–1955). Alongside the insights of mathematicians like Hendrik Lorentz (1853–1928), David Hilbert (1862–1943), Hermann Minkowski (1864–1909), and Emmy Noether (1882–1935), among others, Einstein posited that reality was not framed by absolute space or absolute time, but was *relative*, which is to say, subject to local measurements of a single four-dimensional manifold: space-time. For Einstein, however, the fourth dimension was not a hidden spatial dimension, but time. Moreover, this space-time was in fact curved, warped by the gravitational effects of massive objects. This meant that rather than simply a bizarre off-shoot of regular geometry with little real-world application, the universe itself was fundamentally non-Euclidean, and could not be understood otherwise.

As with Hinton's *Scientific Romances*, attempts were made to explain these new concepts through the medium of fiction, the most influential of these being George Gamow's *Mr. Tompkins in Wonderland* (1940). It was in the annals of weird and science fiction, however, that Einsteinian physics would find its most imaginative expression. This was also the era of a revolutionary new branch of physics known as quantum mechanics which offered up concepts like "entanglement", or, as Einstein put it denigratingly, "spooky actions at a distance".* That higher dimensional geometry, relativity, and quantum theory came to fruition in a period of only forty years or so meant that their uses in fiction were often confused, if not wildly inaccurate. Indeed, the very complexity of these ideas

* Einstein, Albert. "84, Letter to Max Born, 3 March, 1947". *The Born-Einstein Letters.* Trans. Irene Born. London: Macmillan, 1971. 158.

gave authors a kind of *carte blanche* to conjure all sorts of "Amazing" and "Astounding" fictions, as the titles of the pulp magazines they appeared in would so memorably phrase it.

The influence of quantum mechanics on weird fiction intensified as the century progressed. The field's initial foundations were determined at the 1927 Solvay Conference in Belgium, a conference attended by such luminaries of mathematics as Einstein, Max Planck, and Werner Heisenberg, among many others. These discoveries and debates proved fertile material for writers of the weird and the supernatural, who often bent quantum concepts along more mystical lines. In the feverish short story "The Dreams in the Witch-House" (1933), for example, H. P. Lovecraft's mathematician protagonist, Walter Gilman, cites these three figures by name, only to make discoveries beyond even their wildest dreams. This story—one of Lovecraft's most famous—is not included in this particular volume as it has been reprinted extensively over the years including in the British Library's own *The Gothic Tales of H. P. Lovecraft* (2018). For this reason, a more obscure story, co-authored with Henry S. Whitehead, has been selected instead. The most influential and notorious of all weird fiction writers, Lovecraft's very prose seems non-Euclidean at times, bending and contorting in unexpected ways.

To bring this introduction to a close, I end with a confession and a disclaimer. The confession: I am not a mathematician. Like many, however, I have long been drawn to the power and ingenuity of the discipline, and marvel at the role it has played in so many of humanity's greatest achievements, and the role it continues to play so invisibly and yet so fundamentally in our everyday lives. Thus, the disclaimer: this is not a book of mathematical problems or puzzles.

At least, not in any straightforward sense. In this, the card-carrying mathematician may be disappointed. The short story, it must be noted, is not an especially apt form for mathematics of any kind. Spending a page or even a few paragraphs explaining complex mathematical phenomena does not a compelling story make, and such narratives are rarely effective (though there are, of course, exceptions like Arthur Porges' comedic 1954 tale, "The Devil and Simon Flagg", where solving Fermat's Last Theorem proves a bridge too far for the Devil). As such, the present collection has been governed throughout by a requirement that the stories in this volume should be just that—stories, not mathematical treatises clothed in a thin raiment of fiction. They are all imaginative, speculative, affecting; they are all, to use Wilde's expression, poetical.

As for higher dimensions, the twenty-first century has not done away with them just yet. On the contrary, certain factions of contemporary theoretical physics cannot do without them. M-Theory, for example, posits eleven dimensions, while Bosonic String Theory requires a mind-boggling twenty-six. Reality may yet prove stranger than fiction. To paraphrase the cultural theorist Mark Fisher—black holes are weirder than vampires.* Looking back to the weird fiction of the late nineteenth and early twentieth century, however, we find a strangely amorphous genre, its gelatinous body digesting elements of the Gothic, the ghost story, sensation fiction, science fiction, and even detective fiction, without submitting itself to any one set of established rules or conventions. The stories presented here waver between genres, but they are all of them strange. It is only fitting, then, that the last words should go to the narrator of

* Fisher, Mark. *The Weird and the Eerie*. London: Repeater Books, 2016. 15.

what is perhaps Lovecraft's strangest tale of all, "The Dreams in the Witch-House":

> "Non-Euclidean calculus and quantum physics are enough to stretch any brain; and when one mixes them with folklore, and tries to trace a strange background of multi-dimensional reality behind the ghoulish hints of the Gothic tales and the wild whispers of the chimney-corner, one can hardly expect to be wholly free from mental tension."*

HENRY BARTHOLOMEW

DR HENRY BARTHOLOMEW is a teacher and researcher specializing in literary theory and late nineteenth- and early twentieth-century supernatural fiction. He has published work on the Gothic, the 'uncanny', and Speculative Realism, and is currently researching the metaphysics of beauty and ecology in the work of several ghost-story writers including Algernon Blackwood and Vernon Lee.

* Lovecraft, H. P. *The Gothic Tales of H. P. Lovecraft*. Ed. Xavier Aldana Reyes. London: British Library Publishing, 2018. 160.

FURTHER READING

Fiction

Fantasia Mathematica: Being a Set of Stories, Together with a Group of Oddments and Diversions, All Drawn from the Universe of Mathematics. Ed. Clifton Fadiman. New York: Copernicus, 1997.

Mathenauts: Tales of Mathematical Wonder. Ed. Rudy Rucker. Arbor House Publishing, 1987.

Non-Fiction

Blacklock, Mark. "Higher Spatial Form in Weird Fiction". *Textual Practice* (31:6), 2017. 1101–1116.

Henderson, Linda Dalrymple. *The Fourth Dimension and Non-Euclidean Geometry in Modern Art.* Princeton University Press, 1983.

Killeen, Jarlath. "Oscar Wilde in the Fourth Dimension: Ghosts, Geometry, and the Victorian Crisis of Meaning". *The Routledge Handbook to the Ghost Story.* Ed. Scott Brewster and Luke Thurston. New York: Routledge, 2018. 49–58.

Throesch, Elizabeth L. *Before Einstein: The Fourth Dimension in Fin-de-Siècle Literature and Culture.* London: Anthem Press, 2017.

ACKNOWLEDGEMENTS

I would like to thank Jonny Davidson and the British Library publishing team for their help and encouragement in bringing this strange project into being. Likewise, my thanks and admiration are extended to Sandra Gómez and Mauricio Villamayor for their striking cover illustration and design work. Thanks are due, also, to Mike Ashley for suggesting the inclusion of Miriam Allen deFord's excellent story "Slips Take Over", and for his decades of scholarship on all things strange and speculative. Finally, a debt of gratitude is owed to Nick Groom and Eleanor Hopkins for their insightful comments on early drafts, and to Mitchell Berger for his generous feedback on the mathematical elements of the introduction.

—1896—

THE PLATTNER STORY

H. G. Wells

H. G. Wells (1866–1946) needs little introduction. Author, futurist, and social commentator, Wells penned some of the greatest science fiction stories of all time, including The Island of Doctor Moreau *(1896),* The War of the Worlds *(1897), and* The Invisible Man *(1897). Two other stories, both from 1895—*The Time Machine *and "The Remarkable Case of Davidson's Eyes"—make use of extra dimensions; the former, remarkably, pre-empts Einstein by explicitly identifying time as the fourth dimension. This little-reprinted story builds on these fictions and is one of the finest early examples of the fourth dimension in supernatural fiction. First published in* New Review *in 1896 and collected in* The Plattner Story and Others *(1897), it is by turns comic, eerie, and profound.*

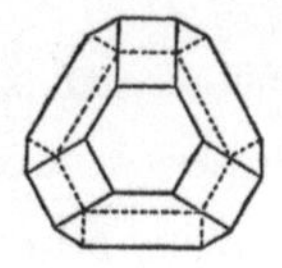

WHETHER THE STORY OF GOTTFRIED PLATTNER IS TO BE credited or not, is a pretty question in the value of evidence. On the one hand, we have seven witnesses—to be perfectly exact, we have six and a half pairs of eyes, and one undeniable fact; and on the other we have—what is it?—prejudice, common sense, the inertia of opinion. Never were there seven more honest-seeming witnesses; never was there a more undeniable fact than the inversion of Gottfried Plattner's anatomical structure, and—never was there a more preposterous story than the one they have to tell! The most preposterous part of the story is the worthy Gottfried's contribution (for I count him as one of the seven). Heaven forbid that I should be led into giving countenance to superstition by a passion for impartiality, and so come to share the fate of Eusapia's patrons! Frankly, I believe there is something crooked about this business of Gottfried Plattner; but what that crooked factor is, I will admit as frankly, I do not know. I have been surprised at the credit accorded to the story in the most unexpected and authoritative quarters. The fairest way to the reader, however, will be for me to tell it without further comment.

Gottfried Plattner is, in spite of his name, a freeborn Englishman. His father was an Alsatian who came to England in the Sixties, married a respectable English girl of unexceptionable antecedents, and died, after a wholesome and uneventful life (devoted, I understand, chiefly to the laying of parquet flooring), in 1887. Gottfried's age is seven-and-twenty. He is, by virtue of his heritage of three languages,

Modern Languages Master in a small private school in the South of England. To the casual observer he is singularly like any other Modern Languages Master in any other small private school. His costume is neither very costly nor very fashionable, but, on the other hand, it is not markedly cheap or shabby; his complexion, like his height and his bearing, is inconspicuous. You would notice, perhaps, that, like the majority of people, his face was not absolutely symmetrical, his right eye a little larger than the left, and his jaw a trifle heavier on the right side. If you, as an ordinary careless person, were to bare his chest and feel his heart beating, you would probably find it quite like the heart of anyone else. But here you and the trained observer would part company. If you found his heart quite ordinary, the trained observer would find it quite otherwise. And once the thing was pointed out to you, you too would perceive the peculiarity easily enough. It is that Gottfried's heart beats on the right side of his body.

Now, that is not the only singularity of Gottfried's structure, although it is the only one that would appeal to the untrained mind. Careful sounding of Gottfried's internal arrangements, by a well-known surgeon, seems to point to the fact that all the other unsymmetrical parts of his body are similarly misplaced. The right lobe of his liver is on the left side, the left on his right; while his lungs, too, are similarly contraposed. What is still more singular, unless Gottfried is a consummate actor, we must believe that his right hand has recently become his left. Since the occurrences we are about to consider (as impartially as possible), he has found the utmost difficulty in writing, except from right to left across the paper with his left hand. He cannot throw with his right hand, he is perplexed at meal times between knife and fork, and his ideas of the rule of the road—he is a cyclist—are still a dangerous confusion. And there is not a scrap of evidence to show that before these occurrences Gottfried was at all left-handed.

There is yet another wonderful fact in this preposterous business. Gottfried produces three photographs of himself. You have him at the age of five or six, thrusting fat legs at you from under a plaid frock, and scowling. In that photograph his left eye is a little larger than his right, and his jaw is a trifle heavier on the left side. This is the reverse of his present living conditions. The photograph of Gottfried at fourteen seems to contradict these facts, but that is because it is one of those cheap "Gem" photographs that were then in vogue, taken direct upon metal, and therefore reversing things just as a looking-glass would. The third photograph represents him at one-and-twenty, and confirms the record of the others. There seems here evidence of the strongest confirmatory character that Gottfried has exchanged his left side for his right. Yet how a human being can be so changed, short of a fantastic and pointless miracle, it is exceedingly hard to suggest.

In one way, of course, these facts might be explicable on the supposition that Plattner has undertaken an elaborate mystification, on the strength of his heart's displacement. Photographs may be fudged, and left-handedness imitated. But the character of the man does not lend itself to any such theory. He is quiet, practical, unobtrusive, and thoroughly sane, from the Nordau standpoint. He likes beer, and smokes moderately, takes walking exercise daily, and has a healthily high estimate of the value of his teaching. He has a good but untrained tenor voice, and takes a pleasure in singing airs of a popular and cheerful character. He is fond, but not morbidly fond, of reading,—chiefly fiction pervaded with a vaguely pious optimism,—sleeps well, and rarely dreams. He is, in fact, the very last person to evolve a fantastic fable. Indeed, so far from forcing this story upon the world, he has been singularly reticent on the matter. He meets inquirers with a certain engaging—bashfulness is almost the word, that disarms the most suspicious. He seems genuinely ashamed that anything so unusual has occurred to him.

It is to be regretted that Plattner's aversion to the idea of post-mortem dissection may postpone, perhaps for ever, the positive proof that his entire body has had its left and right sides transposed. Upon that fact mainly the credibility of his story hangs. There is no way of taking a man and moving him about *in space*, as ordinary people understand space, that will result in our changing his sides. Whatever you do, his right is still his right, his left his left. You can do that with a perfectly thin and flat thing, of course. If you were to cut a figure out of paper, any figure with a right and left side, you could change its sides simply by lifting it up and turning it over. But with a solid it is different. Mathematical theorists tell us that the only way in which the right and left sides of a solid body can be changed is by taking that body clean out of space as we know it,—taking it out of ordinary existence, that is, and turning it somewhere outside space. This is a little abstruse, no doubt, but anyone with any knowledge of mathematical theory will assure the reader of its truth. To put the thing in technical language, the curious inversion of Plattner's right and left sides is proof that he has moved out of our space into what is called the Fourth Dimension, and that he has returned again to our world. Unless we choose to consider ourselves the victims of an elaborate and motiveless fabrication, we are almost bound to believe that this has occurred.

So much for the tangible facts. We come now to the account of the phenomena that attended his temporary disappearance from the world. It appears that in the Sussexville Proprietary School, Plattner not only discharged the duties of Modern Languages Master, but also taught chemistry, commercial geography, book-keeping, shorthand, drawing, and any other additional subject to which the changing fancies of the boys' parents might direct attention. He knew little or nothing of these various subjects, but in secondary as distinguished from Board or elementary schools, knowledge in the teacher is, very properly, by no means so necessary as high moral character and gentlemanly tone. In chemistry he was particularly deficient, knowing, he says, nothing beyond the Three Gases (whatever the three gases may be). As, however, his pupils began by knowing nothing, and derived all their information from him, this caused him (or anyone) but little inconvenience for several terms. Then a little boy named Whibble joined the school, who had been educated (it seems) by some mischievous relative into an inquiring habit of mind. This little boy followed Plattner's lessons with marked and sustained interest, and in order to exhibit his zeal on the subject, brought, at various times, substances for Plattner to analyse. Plattner, flattered by this evidence of his power of awakening interest, and trusting to the boy's ignorance, analysed these, and even made general statements as to their composition. Indeed, he was so far stimulated by his pupil as to obtain a work upon analytical chemistry, and study it during his supervision of the evening's preparation. He was surprised to find chemistry quite an interesting subject.

So far the story is absolutely commonplace. But now the greenish powder comes upon the scene. The source of that greenish powder seems, unfortunately, lost. Master Whibble tells a tortuous story of finding it done up in a packet in a disused limekiln near the Downs. It would have been an excellent thing for Plattner, and possibly for Master Whibble's family, if a match could have been applied to that powder there and then. The young gentleman certainly did not bring it to school in a packet, but in a common eight-ounce graduated medicine bottle, plugged with masticated newspaper. He gave it to Plattner at the end of the afternoon school. Four boys had been detained after school prayers in order to complete some neglected tasks, and Plattner was supervising these in the small classroom in which the chemical teaching was conducted. The appliances for the practical teaching of chemistry in the Sussexville Proprietary School, as in most small schools in this country, are characterized by a severe simplicity. They are kept in a small cupboard standing in a recess, and having about the same capacity as a common travelling trunk. Plattner, being bored with his passive superintendence, seems to have welcomed the intervention of Whibble with his green powder as an agreeable diversion, and, unlocking this cupboard, proceeded at once with his analytical experiments. Whibble sat, luckily for himself, at a safe distance, regarding him. The four malefactors, feigning a profound absorption in their work, watched him furtively with the keenest interest. For even within the limits of the Three Gases, Plattner's practical chemistry was, I understand, temerarious.

They are practically unanimous in their account of Plattner's proceedings. He poured a little of the green powder into a test-tube, and tried the substance with water, hydrochloric acid, nitric acid, and sulphuric acid in succession. Getting no result, he emptied out a little heap—nearly half the bottleful, in fact—upon a slate and tried a match. He held the medicine bottle in his left hand. The stuff began to smoke and melt, and then—exploded with deafening violence and a blinding flash.

The five boys, seeing the flash and being prepared for catastrophes, ducked below their desks, and were none of them seriously hurt. The window was blown out into the playground, and the blackboard on its easel was upset. The slate was smashed to atoms. Some plaster fell from the ceiling. No other damage was done to the school edifice or appliances, and the boys at first, seeing nothing of Plattner, fancied he was knocked down and lying out of their sight below the desks. They jumped out of their places to go to his assistance, and were amazed to find the space empty. Being still confused by the sudden violence of the report, they hurried to the open door, under the impression that he must have been hurt, and have rushed out of the room. But Carson, the foremost, nearly collided in the doorway with the principal, Mr. Lidgett.

Mr. Lidgett is a corpulent, excitable man with one eye. The boys describe him as stumbling into the room mouthing some of those tempered expletives irritable schoolmasters accustom themselves to use—lest worse befall. "Wretched mumchancer!" he said. "Where's Mr. Plattner?" The boys are agreed on the very words. ("Wobbler," "snivelling puppy," and "mumchancer" are, it seems, among the ordinary small change of Mr. Lidgett's scholastic commerce.)

Where's Mr. Plattner? That was a question that was to be repeated many times in the next few days. It really seemed as though that frantic hyperbole, "blown to atoms," had for once realized itself. There was not a visible particle of Plattner to be seen; not a drop of blood nor a stitch of clothing to be found. Apparently he had been blown clean out of existence and left not a wrack behind.

Not so much as would cover a sixpenny piece, to quote a proverbial expression! The evidence of his absolute disappearance, as a consequence of that explosion, is indubitable.

It is not necessary to enlarge here upon the commotion excited in the Sussexville Proprietary School, and in Sussexville and elsewhere, by this event. It is quite possible, indeed, that some of the readers of these pages may recall the hearing of some remote and dying version of that excitement during the last summer holidays. Lidgett, it would seem, did everything in his power to suppress and minimize the story. He instituted a penalty of twenty-five lines for any mention of Plattner's name among the boys, and stated in the schoolroom that he was clearly aware of his assistant's whereabouts. He was afraid, he explains, that the possibility of an explosion happening, in spite of the elaborate precautions taken to minimize the practical teaching of chemistry, might injure the reputation of the school; and so might any mysterious quality in Plattner's departure. Indeed, he did everything in his power to make the occurrence seem as ordinary as possible. In particular, he cross-examined the five eye-witnesses of the occurrence so searchingly that they began to doubt the plain evidence of their senses. But, in spite of these efforts, the tale, in a magnified and distorted state, made a nine days' wonder in the district, and several parents withdrew their sons on colourable pretexts. Not the least remarkable point in the matter is the fact that a large number of people in the neighbourhood dreamed singularly vivid dreams of Plattner during the period of excitement before his return, and that these dreams had a curious uniformity. In almost all of them Plattner was seen, sometimes singly, sometimes in company, wandering about through a coruscating iridescence. In all cases his face was pale and distressed, and in some he gesticulated towards the dreamer. One or two of the boys, evidently under the influence of nightmare, fancied that Plattner approached them with remarkable swiftness, and seemed to look closely into their very eyes. Others fled with Plattner from the pursuit of vague and extraordinary creatures of a globular shape. But all these fancies were forgotten in inquiries and speculations when, on the Wednesday next but one after the Monday of the explosion, Plattner returned.

The circumstances of his return were as singular as those of his departure. So far as Mr. Lidgett's somewhat choleric outline can be filled in from Plattner's hesitating statements, it would appear that on Wednesday evening, towards the hour of sunset, the former gentleman, having dismissed evening preparation, was engaged in his garden, picking and eating strawberries, a fruit of which he is inordinately fond. It is a large old-fashioned garden, secured from observation, fortunately, by a high and ivy-covered red-brick wall. Just as he was stooping over a particularly prolific plant, there was a flash in the air and a heavy thud, and before he could look round, some heavy body struck him violently from behind. He was pitched forward, crushing the strawberries he held in his hand, and that so roughly, that his silk hat—Mr. Lidgett adheres to the older ideas of scholastic costume—was driven violently down upon his forehead, and almost over one eye. This heavy missile, which slid over him sideways and collapsed into a sitting posture among the strawberry plants, proved to be our long-lost Mr. Gottfried Plattner, in an extremely dishevelled condition. He was collarless and hatless, his linen was dirty, and there was blood upon his hands. Mr. Lidgett was so indignant and surprised that he remained on all-fours, and with his hat jammed down on his eye, while he expostulated vehemently with Plattner for his disrespectful and unaccountable conduct.

This scarcely idyllic scene completes what I may call the exterior version of the Plattner story—its exoteric aspect. It is quite unnecessary to enter here into all the details of his dismissal by Mr. Lidgett. Such details, with the full names and dates and references, will be found in the larger report of these occurrences that was laid before the Society for the Investigation of Abnormal Phenomena. The singular transposition of Plattner's right and left sides was scarcely observed for the first day or so, and then first in connection with his disposition to write from right to left across the blackboard. He concealed rather than ostended this curious confirmatory circumstance, as he considered it would unfavourably affect his prospects in a new situation. The displacement of his heart was discovered some months after, when he was having a tooth extracted under anæsthetics. He then, very unwillingly, allowed a cursory surgical examination to be made of himself, with a view to a brief account in the *Journal of Anatomy*. That exhausts the statement of the material facts; and we may now go on to consider Plattner's account of the matter.

But first let us clearly differentiate between the preceding portion of this story and what is to follow. All I have told thus far is established by such evidence as even a criminal lawyer would approve. Every one of the witnesses is still alive; the reader, if he have the leisure, may hunt the lads out tomorrow, or even brave the terrors of the redoubtable Lidgett, and cross-examine and trap and test to his heart's content; Gottfried Plattner, himself, and his twisted heart and his three photographs are producible. It may be taken as proved that he did disappear for nine days as the consequence of an explosion; that he returned almost as violently, under circumstances in their nature annoying to Mr. Lidgett, whatever the details of those circumstances may be; and that he returned inverted, just as a reflection returns from a mirror. From the last fact, as I have already stated, it follows almost inevitably that Plattner, during those nine days, must have been in some state of existence altogether out of space. The evidence to these statements is, indeed, far stronger than that upon which most murderers are hanged. But for his own particular account of where he had been, with its confused explanations and well-nigh self-contradictory details, we have only Mr. Gottfried Plattner's word. I do not wish to discredit that, but I must point out—what so many writers upon obscure psychic phenomena fail to do—that we are passing here from the practically undeniable to that kind of matter which any reasonable man is entitled to believe or reject as he thinks proper. The previous statements render it plausible; its discordance with common experience tilts it towards the incredible. I would prefer not to sway the beam of the reader's judgment either way, but simply to tell the story as Plattner told it me.

He gave me his narrative, I may state, at my house at Chislehurst, and so soon as he had left me that evening, I went into my study and wrote down everything as I remembered it. Subsequently he was good enough to read over a type-written copy, so that its substantial correctness is undeniable.

He states that at the moment of the explosion he distinctly thought he was killed. He felt lifted off his feet and driven forcibly backward. It is a curious fact for psychologists that he thought clearly during his backward flight, and wondered whether he should hit the chemistry cupboard or the blackboard easel. His heels struck ground, and he staggered and fell heavily into a sitting position on something soft and firm. For a moment the concussion stunned him. He became aware at once of a vivid scent of singed hair, and he seemed to hear the voice of Lidgett asking for him. You will understand that for a time his mind was greatly confused.

At first he was distinctly under the impression that he was still in the classroom. He perceived quite distinctly the surprise of the boys and the entry of Mr. Lidgett. He is quite positive upon that score. He did not hear their remarks; but that he ascribed to the deafening effect of the experiment. Things about him seemed curiously dark and faint, but his mind explained that on the obvious but mistaken idea that the explosion had engendered a huge volume of dark smoke. Through the dimness the figures of Lidgett and the boys moved, as faint and silent as ghosts. Plattner's face still tingled with the stinging heat of the flash. He was, he says, "all muddled." His first definite thoughts seem to have been of his personal safety. He thought he was perhaps blinded and deafened. He felt his limbs and face in a gingerly manner. Then his perceptions grew clearer, and he was astonished to miss the old familiar desks and other schoolroom furniture about him. Only dim, uncertain, grey shapes stood in the place of these. Then came a thing that made him shout aloud, and awoke his stunned faculties to instant activity. *Two of the boys, gesticulating, walked one after the other clean through him!* Neither manifested the slightest consciousness of his presence. It is difficult to imagine the sensation he felt. They came against him, he says, with no more force than a wisp of mist.

Plattner's first thought after that was that he was dead. Having been brought up with thoroughly sound views in these matters, however, he was a little surprised to find his body still about him.

His second conclusion was that he was not dead, but that the others were: that the explosion had destroyed the Sussexville Proprietary School and every soul in it except himself. But that, too, was scarcely satisfactory. He was thrown back upon astonished observation.

Everything about him was extraordinarily dark: at first it seemed to have an altogether ebony blackness. Overhead was a black firmament. The only touch of light in the scene was a faint greenish glow at the edge of the sky in one direction, which threw into prominence a horizon of undulating black hills. This, I say, was his impression at first. As his eye grew accustomed to the darkness, he began to distinguish a faint quality of differentiating greenish colour in the circumambient night. Against this background the furniture and occupants of the classroom, it seems, stood out like phosphorescent spectres, faint and impalpable. He extended his hand, and thrust it without an effort through the wall of the room by the fireplace.

He describes himself as making a strenuous effort to attract attention. He shouted to Lidgett, and tried to seize the boys as they went to and fro. He only desisted from these attempts when Mrs. Lidgett, whom he (as an Assistant Master) naturally disliked, entered the room. He says the sensation of being in the world, and yet not a part of it, was an extraordinarily disagreeable one. He compared his feelings, not inaptly, to those of a cat watching a mouse through a window. Whenever he made a motion to communicate with the dim, familiar world about him, he found an invisible, incomprehensible barrier preventing intercourse.

He then turned his attention to his solid environment. He found the medicine bottle still unbroken in his hand, with the remainder of the green powder therein. He put this in his pocket, and began to feel about him. Apparently, he was sitting on a boulder of rock covered with a velvety moss. The dark country about him he was unable to see, the faint, misty picture of the schoolroom blotting it out, but he had a feeling (due perhaps to a cold wind) that he was near the crest of a hill, and that a steep valley fell away beneath his feet. The green glow along the edge of the sky seemed to be growing in extent and intensity. He stood up, rubbing his eyes.

It would seem that he made a few steps, going steeply downhill, and then stumbled, nearly fell, and sat down again upon a jagged mass of rock to watch the dawn. He became aware that the world about him was absolutely silent. It was as still as it was dark, and though there was a cold wind blowing up the hill-face, the rustle of grass, the soughing of the boughs that should have accompanied it, were absent. He could hear, therefore, if he could not see, that the hillside upon which he stood was rocky and desolate. The green grew brighter every moment, and as it did so a faint, transparent blood-red mingled with, but did not mitigate, the blackness of the sky overhead and the rocky desolations about him. Having regard to what follows, I am inclined to think that that redness may have been an optical effect due to contrast. Something black fluttered momentarily against the livid yellow-green of the lower sky, and then the thin and penetrating voice of a bell rose out of the black gulf below him. An oppressive expectation grew with the growing light.

It is probable that an hour or more elapsed while he sat there, the strange green light growing brighter every moment, and spreading slowly, in flamboyant fingers, upward towards the zenith. As it grew, the spectral vision of *our* world became relatively or absolutely fainter. Probably both, for the time must have been about that of our earthly sunset. So far as his vision of our world went, Plattner, by his few steps downhill, had passed through the floor of the classroom, and was now, it seemed, sitting in mid-air in the larger schoolroom downstairs. He saw the boarders distinctly, but much more faintly than he had seen Lidgett. They were preparing their evening tasks, and he noticed with interest that several were cheating with their Euclid riders by means of a crib, a compilation whose existence he had hitherto never suspected. As the time passed, they faded steadily, as steadily as the light of the green dawn increased.

Looking down into the valley, he saw that the light had crept far down its rocky sides, and that the profound blackness of the abyss was now broken by a minute green glow, like the light of a glow-worm. And almost immediately the limb of a huge heavenly body of blazing green rose over the basaltic undulations of the distant hills, and the monstrous hill-masses about him came out gaunt and desolate, in green light and deep, ruddy black shadows. He became aware of a vast number of ball-shaped objects drifting as thistledown drifts over the high ground. There were none of these nearer to him than the opposite side of the gorge. The bell below twanged quicker and quicker, with something like impatient insistence, and several lights moved hither and thither. The boys at work at their desks were now almost imperceptibly faint.

This extinction of our world, when the green sun of this other universe rose, is a curious point upon which Plattner insists. During the Other-World night it is difficult to move about, on account of the vividness with which the things of this world are visible. It becomes a riddle to explain why, if this is the case, we in this world catch no glimpse of the Other-World. It is due, perhaps, to the comparatively vivid illumination of this world of ours. Plattner describes the midday of the Other-World, at its brightest, as not being nearly so bright as this world at full moon, while its night is profoundly black. Consequently, the amount of light, even in an ordinary dark room, is sufficient to render the things of the Other-World invisible, on the same principle that faint phosphorescence is only visible in the profoundest darkness. I have tried, since he told me his story, to see something of the Other-World by sitting for a long space in a photographer's dark room at night. I have certainly seen indistinctly the form of greenish slopes and rocks, but only, I must admit, very indistinctly indeed. The reader may possibly be more successful. Plattner tells me that since his return he has dreamt and seen and recognized places in the Other-World, but this is probably due to his memory of these scenes. It seems quite possible that people with unusually keen eyesight may occasionally catch a glimpse of this strange Other-World about us.

However, this is a digression. As the green sun rose, a long street of black buildings became perceptible, though only darkly and indistinctly, in the gorge, and, after some hesitation, Plattner began to clamber down the precipitous descent towards them. The descent was long and exceedingly tedious, being so not only by the extraordinary steepness, but also by reason of the looseness of the boulders with which the whole face of the hill was strewn. The noise of his descent—now and then his heels struck fire from the rocks—seemed now the only sound in the universe, for the beating of the bell had ceased. As he drew nearer, he perceived that the various edifices had a singular resemblance to tombs and mausoleums and monuments, saving only that they were all uniformly black instead of being white, as most sepulchres are. And then he saw, crowding out of the largest building, very much as people disperse from church, a number of pallid, rounded, pale-green figures. These dispersed in several directions about the broad street of the place, some going through side alleys and reappearing upon the steepness of the hill, others entering some of the small black buildings which lined the way.

At the sight of these things drifting up towards him, Plattner stopped, staring. They were not walking, they were indeed limbless, and they had the appearance of human heads, beneath which a tadpole-like body swung. He was too astonished at their strangeness, too full, indeed, of strangeness, to be seriously alarmed by them. They drove towards him, in front of the chill wind that was blowing uphill, much as soap-bubbles drive before a draught. And as he looked at the nearest of those approaching, he saw it was indeed a human head, albeit with singularly large eyes, and wearing such an expression of distress and anguish as he had never seen before upon mortal countenance. He was surprised to find that it did not turn to regard him, but seemed to be watching and following some unseen moving thing. For a moment he was puzzled, and then it occurred to him that this creature was watching with its enormous eyes something that was happening in the world he had just left. Nearer it came, and nearer, and he was too astonished to cry out. It made a very faint fretting sound as it came close to him. Then it struck his face with a gentle pat—its touch was very cold—and drove past him, and upward towards the crest of the hill.

An extraordinary conviction flashed across Plattner's mind that this head had a strong likeness to Lidgett. Then he turned his attention to the other heads that were now swarming thickly, up the hillside. None made the slightest sign of recognition. One or two, indeed, came close to his head and almost followed the example of the first, but he dodged convulsively out of the way. Upon most of them he saw the same expression of unavailing regret he had seen upon the first, and heard the same faint sounds of wretchedness from them. One or two wept, and one rolling swiftly uphill wore an expression of diabolical rage. But others were cold, and several had a look of gratified interest in their eyes. One, at least, was almost in an ecstasy of happiness. Plattner does not remember that he recognized any more likenesses in those he saw at this time.

For several hours, perhaps, Plattner watched these strange things dispersing themselves over the hills, and not till long after they had ceased to issue from the clustering black buildings in the gorge, did he resume his downward climb. The darkness about him increased so much that he had a difficulty in stepping true. Overhead the sky was now a bright, pale green. He felt neither hunger nor thirst. Later, when he did, he found a chilly stream running down the centre of the gorge, and the rare moss upon the boulders, when he tried it at last in desperation, was good to eat.

He groped about among the tombs that ran down the gorge, seeking vaguely for some clue to these inexplicable things. After a long time he came to the entrance of the big mausoleum-like building from which the heads had issued. In this he found a group of green lights burning upon a kind of basaltic altar, and a bell-rope from a belfry overhead hanging down into the centre of the place. Round the wall ran a lettering of fire in a character unknown to him. While he was still wondering at the purport of these things, he heard the receding tramp of heavy feet echoing far down the street. He ran out into the darkness again, but he could see nothing. He had a mind to pull the bell-rope, and finally decided to follow the footsteps. But, although he ran far, he never overtook them; and his shouting was of no avail. The gorge seemed to extend an interminable distance. It was as dark as earthly starlight throughout its length, while the ghastly green day lay along the upper edge of its precipices. There were none of the heads, now, below. They were all, it seemed, busily occupied along the upper slopes. Looking up, he saw them drifting hither and thither, some hovering stationary, some flying swiftly through the air. It reminded him, he said, of "big snowflakes"; only these were black and pale green.

In pursuing the firm, undeviating footsteps that he never overtook, in groping into new regions of this endless devil's dyke, in clambering up and down the pitiless heights, in wandering about the summits, and in watching the drifting faces, Plattner states that he spent the better part of seven or eight days. He did not keep count, he says. Though once or twice he found eyes watching him, he had word with no living soul. He slept among the rocks on the hillside. In the gorge things earthly were invisible, because, from the earthly standpoint, it was far underground. On the altitudes, so soon as the earthly day began, the world became visible to him. He found himself sometimes stumbling over the dark green rocks, or arresting himself on a precipitous brink, while all about him the green branches of the Sussexville lanes were swaying; or, again, he seemed to be walking through the Sussexville streets, or watching unseen the private business of some household. And then it was he discovered, that to almost every human being in our world there pertained some of these drifting heads: that everyone in the world is watched intermittently by these helpless disembodiments.

What are they—these Watchers of the Living? Plattner never learned. But two, that presently found and followed him, were like his childhood's memory of his father and mother. Now and then other faces turned their eyes upon him: eyes like those of dead people who had swayed him, or injured him, or helped him in his youth and manhood. Whenever they looked at him, Plattner was overcome with a strange sense of responsibility. To his mother he ventured to speak; but she made no answer. She looked sadly, steadfastly, and tenderly—a little reproachfully, too, it seemed—into his eyes.

He simply tells this story: he does not endeavour to explain. We are left to surmise who these Watchers of the Living may be, or if they are indeed the Dead, why they should so closely and passionately watch a world they have left for ever. It may be—indeed to my mind it seems just—that, when our life has closed, when evil or good is no longer a choice for us, we may still have to witness the working out of the train of consequences we have laid. If human souls continue after death, then surely human interests continue after death. But that is merely my own guess at the meaning of the things seen. Plattner offers no interpretation, for none was given him. It is well the reader should understand this clearly. Day after day, with his head reeling, he wandered about this strange-lit world outside the world, weary and, towards the end, weak and hungry. By day—by our earthly day, that is—the ghostly vision of the old familiar scenery of Sussexville, all about him, irked and worried him. He could not see where to put his feet, and ever and again with a chilly touch one of these Watching Souls would come against his face. And after dark the multitude of these Watchers about him, and their intent distress, confused his mind beyond describing. A great longing to return to the earthly life that was so near and yet so remote consumed him. The unearthliness of things about him produced a positively painful mental distress. He was worried beyond describing by his own particular followers. He would shout at them to desist from staring at him, scold at them, hurry away from them. They were always mute and intent. Run as he might over the uneven ground, they followed his destinies.

On the ninth day, towards evening, Plattner heard the invisible footsteps approaching, far away down the gorge. He was then wandering over the broad crest of the same hill upon which he had fallen in his entry into this strange Other-World of his. He turned to hurry down into the gorge, feeling his way hastily, and was arrested by the sight of the thing that was happening in a room in a back street near the school. Both of the people in the room he knew by sight. The windows were open, the blinds up, and the setting sun shone clearly into it, so that it came out quite brightly at first, a vivid oblong of room, lying like a magic-lantern picture upon the black landscape and the livid green dawn. In addition to the sunlight, a candle had just been lit in the room.

On the bed lay a lank man, his ghastly white face terrible upon the tumbled pillow. His clenched hands were raised above his head. A little table beside the bed carried a few medicine bottles, some toast and water, and an empty glass. Every now and then the lank man's lips fell apart, to indicate a word he could not articulate. But the woman did not notice that he wanted anything, because she was busy turning out papers from an old-fashioned bureau in the opposite corner of the room. At first the picture was very vivid indeed, but as the green dawn behind it grew brighter and brighter, so it became fainter and more and more transparent.

As the echoing footsteps paced nearer and nearer, those footsteps that sound so loud in that Other-World and come so silently in this, Plattner perceived about him a great multitude of dim faces gathering together out of the darkness and watching the two people in the room. Never before had he seen so many of the Watchers of the Living. A multitude had eyes only for the sufferer in the room, another multitude, in infinite anguish, watched the woman as she hunted with greedy eyes for something she could not find. They crowded about Plattner, they came across his sight and buffeted his face, the noise of their unavailing regrets was all about him. He saw clearly only now and then. At other times the picture quivered dimly, through the veil of green reflections upon their movements. In the room it must have been very still, and Plattner says the candle flame streamed up into a perfectly vertical line of smoke, but in his ears each footfall and its echoes beat like a clap of thunder. And the faces! Two, more particularly near the woman's: one a woman's also, white and clear-featured, a face which might have once been cold and hard, but which was now softened by the touch of a wisdom strange to earth. The other might have been the woman's father. Both were evidently absorbed in the contemplation of some act of hateful meanness, so it seemed, which they could no longer guard against and prevent. Behind were others, teachers, it may be, who had taught ill, friends whose influence had failed. And over the man, too—a multitude, but none that seemed to be parents or teachers! Faces that might once have been coarse, now purged to strength by sorrow! And in the forefront one face, a girlish one, neither angry nor remorseful, but merely patient and weary, and, as it seemed to Plattner, waiting for relief. His powers of description fail him at the memory of this multitude of ghastly countenances. They gathered on the stroke of the bell. He saw them all in the space of a second. It would seem that he was so worked on by his excitement that, quite involuntarily, his restless fingers took the bottle of green powder out of his pocket and held it before him. But he does not remember that.

Abruptly the footsteps ceased. He waited for the next, and there was silence, and then suddenly, cutting through the unexpected stillness like a keen, thin blade, came the first stroke of the bell. At that the multitudinous faces swayed to and fro, and a louder crying began all about him. The woman did not hear; she was burning something now in the candle flame. At the second stroke everything grew dim, and a breath of wind, icy cold, blew through the host of watchers. They swirled about him like an eddy of dead leaves in the spring, and at the third stroke something was extended through them to the bed. You have heard of a beam of light. This was like a beam of darkness, and looking again at it, Plattner saw that it was a shadowy arm and hand.

The green sun was now topping the black desolations of the horizon, and the vision of the room was very faint. Plattner could see that the white of the bed struggled, and was convulsed; and that the woman looked round over her shoulder at it, startled.

The cloud of watchers lifted high like a puff of green dust before the wind, and swept swiftly downward towards the temple in the gorge. Then suddenly Plattner understood the meaning of the shadowy black arm that stretched across his shoulder and clutched its prey. He did not dare turn his head to see the Shadow behind the arm. With a violent effort, and covering his eyes, he set himself to run, made, perhaps, twenty strides, then slipped on a boulder, and fell. He fell forward on his hands; and the bottle smashed and exploded as he touched the ground.

In another moment he found himself, stunned and bleeding, sitting face to face with Lidgett in the old walled garden behind the school.

⋆

There the story of Plattner's experiences ends. I have resisted, I believe successfully, the natural disposition of a writer of fiction to dress up incidents of this sort. I have told the thing as far as possible in the order in which Plattner told it to me. I have carefully avoided any attempt at style, effect, or construction. It would have been easy, for instance, to have worked the scene of the death-bed into a kind of plot in which Plattner might have been involved. But, quite apart from the objectionableness of falsifying a most extraordinary true story, any such trite devices would spoil, to my mind, the peculiar effect of this dark world, with its livid green illumination and its drifting Watchers of the Living, which, unseen and unapproachable to us, is yet lying all about us.

It remains to add, that a death did actually occur in Vincent Terrace, just beyond the school garden, and, so far as can be proved, at the moment of Plattner's return. Deceased was a rate-collector and insurance agent. His widow, who was much younger than himself, married last month a Mr. Whymper, a veterinary surgeon of Allbeeding. As the portion of this story given here has in various forms circulated orally in Sussexville, she has consented to my use of her name, on condition that I make it distinctly known that she emphatically contradicts every detail of Plattner's account of her husband's last moments. She burnt no will, she says, although Plattner never accused her of doing so: her husband made but one will, and that just after their marriage. Certainly, from a man who had never seen it, Plattner's account of the furniture of the room was curiously accurate.

One other thing, even at the risk of an irksome repetition, I must insist upon, lest I seem to favour the credulous superstitious view. Plattner's absence from the world for nine days is, I think, proved.

But that does not prove his story. It is quite conceivable that even outside space hallucinations may be possible. That, at least, the reader must bear distinctly in mind.

—1903—

THE HALL BEDROOM

Mary Eleanor Wilkins Freeman

Mary Eleanor Wilkins Freeman (1852–1930) is remembered principally for her many novels and story collections on New England life, and for her close attention to female identity and the social and domestic constraints of womanhood. Her most famous supernatural story is the claustrophobic tale of psychic vampirism, "Luella Miller" (1902). Another early story, "An Old Arithmetician" (1885), is notable for its treatment of an older woman's interest in mathematics, but contains no elements of the weird or the supernatural. By contrast, this story, first published in Collier's *magazine, March 1903, is a veritable feast for the senses, and deftly invokes a sense of the other-worldly.*

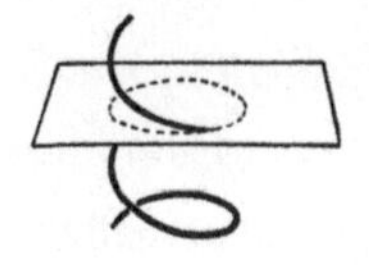

MY NAME IS MRS. ELIZABETH JENNINGS. I AM A HIGHLY respectable woman. I may style myself a gentlewoman, for in my youth I enjoyed advantages. I was well brought up, and I graduated at a young ladies' seminary. I also married well. My husband was that most genteel of all merchants, an apothecary. His shop was on the corner of the main street in Rockton, the town where I was born, and where I lived until the death of my husband. My parents had died when I had been married a short time, so I was left quite alone in the world. I was not competent to carry on the apothecary business by myself, for I had no knowledge of drugs, and had a mortal terror of giving poisons instead of medicines. Therefore I was obliged to sell at a considerable sacrifice, and the proceeds, some five thousand dollars, were all I had in the world. The income was not enough to support me in any kind of comfort, and I saw that I must in some way earn money. I thought at first of teaching, but I was no longer young, and methods had changed since my school days. What I was able to teach, nobody wished to know. I could think of only one thing to do: take boarders. But the same objection to that business as to teaching held good in Rockton. Nobody wished to board. My husband had rented a house with a number of bedrooms, and I advertised, but nobody applied. Finally my cash was running very low, and I became desperate. I packed up my furniture, rented a large house in this town and moved here. It was a venture attended with many risks. In the first place the rent was exorbitant, in the next I was entirely unknown. However, I am a person of considerable ingenuity, and have inventive power, and much enterprise when the occasion presses. I advertised in a very original manner, although that actually took my last penny, that is, the last penny of my ready money, and I was forced to draw on my principal to purchase my first supplies, a thing which I had resolved never on any account to do. But the great risk met with a reward, for I had several applicants within two days after my advertisement appeared in the paper. Within two weeks my boarding-house was well established, I became very successful, and my success would have been uninterrupted had it not been for the mysterious and bewildering occurrences which I am about to relate. I am now forced to leave the house and rent another. Some of my old boarders accompany me, some, with the most unreasonable nervousness, refuse to be longer associated in any way, however indirectly, with the terrible and uncanny happenings which I have to relate. It remains to be seen whether my ill luck in this house will follow me into another, and whether my whole prosperity in life will be forever shadowed by the Mystery of the Hall Bedroom. Instead of telling the strange story myself in my own words, I shall present the Journal of Mr. George H. Wheatcroft. I shall show you the portions beginning on January 18 of the present year, the date when he took up his residence with me. Here it is:

"January 18, 1883. Here I am established in my new boarding-house. I have, as befits my humble means, the hall bedroom, even the hall bedroom on the third floor. I have heard all my life of hall bedrooms, I have seen hall bedrooms, I have been in them, but never until now, when I am actually established in one, did I comprehend what, at once, an ignominious and sternly uncompromising thing a hall bedroom is. It proves the ignominy of the dweller therein. No man at thirty-six (my age) would be domiciled in a hall bedroom, unless he were himself ignominious, at least comparatively speaking. I am proved by this means incontrovertibly to have been left far behind in the race. I see no reason why I should not live in this hall bedroom for the rest of my life, that is, if I have money enough to pay the landlady, and that seems probable, since my small funds are invested as safely as if I were an orphan-ward in charge of a pillar of a sanctuary. After the valuables have been stolen, I have most carefully locked the stable door. I have experienced the revulsion which comes sooner or later to the adventurous soul who experiences nothing but defeat and so-called ill luck. I have swung to the opposite extreme. I have lost in everything—I have lost in love, I have lost in money, I have lost in the struggle for preferment, I have lost in health and strength. I am now settled down in a hall bedroom to live upon my small income, and regain my health by mild potations of the mineral waters here, if possible; if not, to live here without my health—for mine is not a necessarily fatal malady—until Providence shall take me out of my hall bedroom. There is no one place more than another where I care to live. There is not sufficient motive to take me away, even if the mineral waters do not benefit me. So I am here and to stay in the hall bedroom. The landlady is civil, and even kind, as kind as a woman who has to keep her poor womanly eye upon the main chance can be. The struggle for money always injures the fine grain of a woman; she is too fine a thing to do it; she does not by nature belong with the gold grubbers, and it therefore lowers her; she steps from heights to claw and scrape and dig. But she can not help it oftentimes, poor thing, and her deterioration thereby is to be condoned. The landlady is all she can be, taking her strain of adverse circumstances into consideration, and the table is good, even conscientiously so. It looks to me as if she were foolish enough to strive to give the boarders their money's worth, with the due regard for the main chance which is inevitable. However, that is of minor importance to me, since my diet is restricted.

"It is curious what an annoyance a restriction in diet can be even to a man who has considered himself somewhat indifferent to gastronomic delights. There was today a pudding for dinner, which I could not taste without penalty, but which I longed for. It was only because it looked unlike any other pudding that I had ever seen, and assumed a mental and spiritual significance. It seemed to me, whimsically no doubt, as if tasting it might give me a new sensation, and consequently a new outlook. Trivial things may lead to large results: why should I not get a new outlook by means of a pudding? Life here stretches before me most monotonously, and I feel like clutching at alleviations, though paradoxically, since I have settled down with the utmost acquiescence. Still one can not immediately overcome and change radically all one's nature. Now I look at myself critically and search for the keynote to my whole self, and my actions, I have always been conscious of a reaching out, an overweening desire for the new, the untried, for the broadness of further horizons, the seas beyond seas, the thought beyond thought. This characteristic has been the primary cause of all my misfortunes. I have the soul of an explorer, and in nine out of ten cases this leads to destruction. If I had possessed capital and sufficient push, I should have been one of the searchers after the North Pole. I have been an eager student of astronomy. I have studied botany with avidity, and have dreamed of new flora in unexplored parts of the world, and the same with animal life and geology. I longed for riches in order to discover the power and sense of possession of the rich. I longed for love in order to discover the possibilities of the emotions. I longed for all that the mind of man could conceive as desirable for man, not so much for purely selfish ends, as from an insatiable thirst for knowledge of a universal trend. But I have limitations, I do not quite understand of what nature—for what mortal ever did quite understand his own limitations, since a knowledge of them would preclude their existence?—but they have prevented my progress to any extent. Therefore behold me in my hall bedroom, settled at last into a groove of fate so deep that I have lost the sight of even my horizons. Just at present, as I write here, my horizon on the left, that is my physical horizon, is a wall covered with cheap paper. The paper is an indeterminate pattern in white and gilt. There are a few photographs of my own hung about, and on the large wall space beside the bed there is a large oil painting which belongs to my landlady. It has a massive tarnished gold frame, and, curiously enough, the painting itself is rather good. I have no idea who the artist could have been. It is of the conventional landscape type in vogue some fifty years since, the type so fondly reproduced in chromos—the winding river with the little boat occupied by a pair of lovers, the cottage nestled among trees on the right shore, the gentle slope of the hills and the church spire in the background—but still it is well done. It gives me the impression of an artist without the slightest originality of design, but much of technique. But for some inexplicable reason the picture frets me. I find myself gazing at it when I do not wish to do so. It seems to compel my attention like some intent face in the room. I shall ask Mrs. Jennings to have it removed. I will hang in its place some photographs which I have in a trunk.

"January 26. I do not write regularly in my journal. I never did. I see no reason why I should. I see no reason why any one should have the slightest sense of duty in such a matter. Some days I have nothing which interests me sufficiently to write out, some days I feel either too ill or too indolent. For four days I have not written, from a mixture of all three reasons. Now, today I both feel like it and I have something to write. Also I am distinctly better than I have been. Perhaps the waters are benefiting me, or the change of air. Or possibly it is something else more subtle. Possibly my mind has seized upon something new, a discovery which causes it to react upon my failing body and serves as a stimulant. All I know is, I feel distinctly better, and am conscious of an acute interest in doing so, which is of late strange to me. I have been rather indifferent, and sometimes have wondered if that were not the cause rather than the result of my state of health. I have been so continually baulked that I have settled into a state of inertia. I lean rather comfortably against my obstacles. After all, the worst of the pain always lies in the struggle. Give up and it is rather pleasant than otherwise. If one did not kick, the pricks would not in the least matter. However, for some reason, for the last few days, I seem to have awakened from my state of quiescence. It means future trouble for me, no doubt, but in the meantime I am not sorry. It began with the picture—the large oil painting. I went to Mrs. Jennings about it yesterday, and she, to my surprise—for I thought it a matter that could be easily arranged—objected to having it removed. Her reasons were two; both simple, both sufficient, especially since I, after all, had no very strong desire either way. It seems that the picture does not belong to her. It hung here when she rented the house. She says if it is removed, a very large and unsightly discolouration of the wall-paper will be exposed, and she does not like to ask for new paper. The owner, an old man, is travelling abroad, the agent is curt, and she has only been in the house a very short time. Then it would mean a sad upheaval of my room, which would disturb me. She also says that there is no place in the house where she can store the picture, and there is not a vacant space in another room for one so large. So I let the picture remain. It really, when I came to think of it, was very immaterial after all. But I got my photographs out of my trunk, and I hung them around the large picture. The wall is almost completely covered. I hung them yesterday afternoon, and last night I repeated a strange experience which I have had in some degree every night since I have been here, but was not sure whether it deserved the name of experience, but was not rather one of those dreams in which one dreams one is awake. But last night it came again, and now I know. There is something very singular about this room. I am very much interested. I will write down for future reference the events of last night. Concerning those of the preceding nights since I have slept in this room, I will simply say that they have been of a similar nature, but, as it were, only the preliminary stages, the prologue to what happened last night.

"I am not depending upon the mineral waters here as the one remedy for my malady, which is sometimes of an acute nature, and indeed constantly threatens me with considerable suffering unless by medicine I can keep it in check. I will say that the medicine which I employ is not of the class commonly known as drugs. It is impossible that it can be held responsible for what I am about to transcribe. My mind last night and every night since I have slept in this room was in an absolutely normal state. I take this medicine, prescribed by the specialist in whose charge I was before coming here, regularly every four hours while awake. As I am never a good sleeper, it follows that I am enabled with no inconvenience to take any medicine during the night with the same regularity as during the day. It is my habit, therefore, to place my bottle and spoon where I can put my hand upon them easily without lighting the gas. Since I have been in this room, I have placed the bottle of medicine upon my dresser at the side of the room opposite the bed. I have done this rather than place it nearer, as once I jostled the bottle and spilled most of the contents, and it is not easy for me to replace it, as it is expensive. Therefore I placed it in security on the dresser, and, indeed, that is but three or four steps from my bed, the room being so small. Last night I wakened as usual, and I knew, since I had fallen asleep about eleven, that it must be in the neighbourhood of three. I wake with almost clock-like regularity and it is never necessary for me to consult my watch.

"I had slept unusually well and without dreams, and I awoke fully at once, with a feeling of refreshment to which I am not accustomed. I immediately got out of bed and began stepping across the room in the direction of my dresser, on which I had set my medicine-bottle and spoon.

"To my utter amazement, the steps which had hitherto sufficed to take me across my room did not suffice to do so. I advanced several paces, and my outstretched hands touched nothing. I stopped and went on again. I was sure that I was moving in a straight direction, and even if I had not been I knew it was impossible to advance in any direction in my tiny apartment without coming into collision either with a wall or a piece of furniture. I continued to walk falteringly, as I have seen people on the stage: a step, then a long falter, then a sliding step. I kept my hands extended; they touched nothing. I stopped again. I had not the least sentiment of fear or consternation. It was rather the very stupefaction of surprise. 'How is this?' seemed thundering in my ears. 'What is this?'

"The room was perfectly dark. There was nowhere any glimmer, as is usually the case, even in a so-called dark room, from the walls, picture-frames, looking-glass or white objects. It was absolute gloom. The house stood in a quiet part of the town. There were many trees about; the electric street lights were extinguished at midnight; there was no moon and the sky was cloudy. I could not distinguish my one window, which I thought strange, even on such a dark night. Finally I changed my plan of motion and turned, as nearly as I could estimate, at right angles. Now, I thought, I must reach soon, if I kept on, my writing-table underneath the window; or, if I am going in the opposite direction, the hall door. I reached neither. I am telling the unvarnished truth when I say that I began to count my steps and carefully measure my paces after that, and I traversed a space clear of furniture at least twenty feet by thirty—a very large apartment. And as I walked I was conscious that my naked feet were pressing something which gave rise to sensations the like of which I had never experienced before. As nearly as I can express it, it was as if my feet pressed something as elastic as air or water, which was in this case unyielding to my weight. It gave me a curious sensation of buoyancy and stimulation. At the same time this surface, if surface be the right name, which I trod, felt cool to my feet with the coolness of vapour or fluidity, seeming to overlap the soles. Finally I stood still; my surprise was at last merging into a measure of consternation. 'Where am I?' I thought. 'What am I going to do?' Stories that I had heard of travellers being taken from their beds and conveyed into strange and dangerous places, Middle

Age stories of the Inquisition flashed through my brain. I knew all the time that for a man who had gone to bed in a commonplace hall bedroom in a very commonplace little town such surmises were highly ridiculous, but it is hard for the human mind to grasp anything but a human explanation of phenomena. Almost anything seemed then, and seems now, more rational than an explanation bordering upon the supernatural, as we understand the supernatural. At last I called, though rather softly, 'What does this mean?' I said quite aloud, 'Where am I? Who is here? Who is doing this? I tell you I will have no such nonsense. Speak, if there is anybody here.' But all was dead silence. Then suddenly a light flashed through the open transom of my door. Somebody had heard me—a man who rooms next door, a decent kind of man, also here for his health. He turned on the gas in the hall and called to me. 'What's the matter?' he asked, in an agitated, trembling voice. He is a nervous fellow.

"Directly, when the light flashed through my transom, I saw that I was in my familiar hall bedroom. I could see everything quite distinctly—my tumbled bed, my writing-table, my dresser, my chair, my little wash-stand, my clothes hanging on a row of pegs, the old picture on the wall. The picture gleamed out with singular distinctness in the light from the transom. The river seemed actually to run and ripple, and the boat to be gliding with the current. I gazed fascinated at it, as I replied to the anxious voice:

"'Nothing is the matter with me,' said I. 'Why?'

"'I thought I heard you speak,' said the man outside. 'I thought maybe you were sick.'

"'No,' I called back. 'I am all right. I am trying to find my medicine in the dark, that's all. I can see now you have lighted the gas.'

"'Nothing is the matter?'

"'No; sorry I disturbed you. Good-night.'

"'Good-night.' Then I heard the man's door shut after a minute's pause. He was evidently not quite satisfied. I took a pull at my medicine-bottle, and got into bed. He had left the hall-gas burning. I did not go to sleep again for some time. Just before I did so, some one, probably Mrs. Jennings, came out in the hall and extinguished the gas. This morning when I awoke everything was as usual in my room. I wonder if I shall have any such experience tonight.

"January 27. I shall write in my journal every day until this draws to some definite issue. Last night my strange experience deepened, as something tells me it will continue to do. I retired quite early, at half-past ten. I took the precaution, on retiring, to place beside my bed, on a chair, a box of safety matches, that I might not be in the dilemma of the night before. I took my medicine on retiring; that made me due to wake at half-past two. I had not fallen asleep directly, but had had certainly three hours of sound, dreamless slumber when I awoke. I lay a few minutes hesitating whether or not to strike a safety match and light my way to the dresser, whereon stood my medicine-bottle. I hesitated, not because I had the least sensation of fear, but because of the same shrinking from a nerve shock that leads one at times to dread the plunge into an icy bath. It seemed much easier to me to strike that match and cross my hall bedroom to my dresser, take my dose, then return quietly to my bed, than to risk the chance of floundering about in some unknown limbo either of fancy or reality.

"At last, however, the spirit of adventure, which has always been such a ruling one for me, conquered. I rose. I took the box of safety matches in my hand, and started on, as I conceived, the straight course for my dresser, about five feet across from my bed.

As before, I travelled and travelled and did not reach it. I advanced with groping hands extended, setting one foot cautiously before the other, but I touched nothing except the indefinite, unnameable surface which my feet pressed. All of a sudden, though, I became aware of something. One of my senses was saluted, nay, more than that, hailed, with imperiousness, and that was, strangely enough, my sense of smell, but in a hitherto unknown fashion. It seemed as if the odour reached my mentality first. I reversed the usual process, which is, as I understand it, like this: the odour when encountered strikes first the olfactory nerve, which transmits the intelligence to the brain. It is as if, to put it rudely, my nose met a rose, and then the nerve belonging to the sense said to my brain, 'Here is a rose.' This time my brain said, 'Here is a rose,' and my sense then recognized it. I say rose, but it was not a rose, that is, not the fragrance of any rose which I had ever known. It was undoubtedly a flower-odour, and rose came perhaps the nearest to it. My mind realized it first with what seemed a leap of rapture. 'What is this delight?' I asked myself. And then the ravishing fragrance smote my sense. I breathed it in and it seemed to feed my thoughts, satisfying some hitherto unknown hunger. Then I took a step further and another fragrance appeared, which I liken to lilies for lack of something better, and then came violets, then mignonette. I can not describe the experience, but it was a sheer delight, a rapture of sublimated sense. I groped further and further, and always into new waves of fragrance. I seemed to be wading breast-high through flower-beds of Paradise, but all the time I touched nothing with my groping hands. At last a sudden giddiness as of surfeit overcame me. I realized that I might be in some unknown peril. I was distinctly afraid. I struck one of my safety matches, and I was in my hall bedroom, midway between my bed and my dresser. I took my dose of medicine and went to bed, and after a while fell asleep and did not wake till morning.

"January 28. Last night I did not take my usual dose of medicine. In these days of new remedies and mysterious results upon certain organizations, it occurred to me to wonder if possibly the drug might have, after all, something to do with my strange experience.

"I did not take my medicine. I put the bottle as usual on my dresser, since I feared if I interrupted further the customary sequence of affairs I might fail to wake. I placed my box of matches on the chair beside the bed. I fell asleep about quarter past eleven o'clock, and I waked when the clock was striking two—a little earlier than my wont. I did not hesitate this time. I rose at once, took my box of matches and proceeded as formerly. I walked what seemed a great space without coming into collision with anything. I kept sniffing for the wonderful fragrances of the night before, but they did not recur. Instead, I was suddenly aware that I was tasting something, some morsel of sweetness hitherto unknown, and, as in the case of the odour, the usual order seemed reversed, and it was as if I tasted it first in my mental consciousness. Then the sweetness rolled under my tongue. I thought involuntarily of 'Sweeter than honey or the honeycomb' of the Scripture. I thought of the Old Testament manna. An ineffable content as of satisfied hunger seized me. I stepped further, and a new savour was upon my palate. And so on. It was never cloying, though of such sharp sweetness that it fairly stung. It was the merging of a material sense into a spiritual one. I said to myself, 'I have lived my life and always have I gone hungry until now.' I could feel my brain act swiftly under the influence of this heavenly food as under a stimulant. Then suddenly I repeated the experience of the night before. I grew dizzy, and an indefinite fear and shrinking were upon me. I struck my safety match and was back in my hall bedroom. I returned to bed, and soon fell asleep. I did not take my medicine. I am resolved not to do so longer. I am feeling much better.

"January 29. Last night to bed as usual, matches in place; fell asleep about eleven and waked at half-past one. I heard the half-hour strike; I am waking earlier and earlier every night. I had not taken my medicine, though it was on the dresser as usual. I again took my match-box in hand and started to cross the room, and, as always, traversed strange spaces, but this night, as seems fated to be the case every night, my experience was different. Last night I neither smelled nor tasted, but I heard—my Lord, I heard! The first sound of which I was conscious was one like the constantly gathering and receding murmur of a river, and it seemed to come from the wall behind my bed where the old picture hangs. Nothing in nature except a river gives that impression of at once advance and retreat. I could not mistake it. On, ever on, came the swelling murmur of the waves, past and ever past they died in the distance. Then I heard above the murmur of the river a song in an unknown tongue which I recognized as being unknown, yet which I understood; but the understanding was in my brain, with no words of interpretation. The song had to do with me, but with me in unknown futures for which I had no images of comparison in the past; yet a sort of ecstasy as of a prophecy of bliss filled my whole consciousness. The song never ceased, but as I moved on I came into new sound-waves. There was the pealing of bells which might have been made of crystal, and might have summoned to the gates of heaven. There was music of strange instruments, great harmonies pierced now and then by small whispers as of love, and it all filled me with a certainty of a future of bliss.

"At last I seemed the centre of a mighty orchestra which constantly deepened and increased until I seemed to feel myself being lifted gently but mightily upon the waves of sound as upon the waves of a sea. Then again the terror and the impulse to flee to my own familiar scenes was upon me. I struck my match and was back in my hall bedroom. I do not see how I sleep at all after such wonders, but sleep I do. I slept dreamlessly until daylight this morning.

"January 30. I heard yesterday something with regard to my hall bedroom which affected me strangely. I can not for the life of me say whether it intimidated me, filled me with the horror of the abnormal, or rather roused to a greater degree my spirit of adventure and discovery. I was down at the Cure, and was sitting on the veranda sipping idly my mineral water, when somebody spoke my name. 'Mr. Wheatcroft?' said the voice politely, interrogatively, somewhat apologetically, as if to provide for a possible mistake in my identity. I turned and saw a gentleman whom I recognized at once. I seldom forget names or faces. He was a Mr. Addison whom I had seen considerable of three years ago at a little summer hotel in the mountains. It was one of those passing acquaintances which signify little one way or the other. If never renewed, you have no regret; if renewed, you accept the renewal with no hesitation. It is in every way negative. But just now, in my feeble, friendless state, the sight of a face which beams with pleased remembrance is rather grateful. I felt distinctly glad to see the man. He sat down beside me. He also had a glass of the water. His health, while not as bad as mine, leaves much to be desired.

"Addison had often been in this town before. He had in fact lived here at one time. He had remained at the Cure three years, taking the waters daily. He therefore knows about all there is to be known about the town, which is not very large. He asked me where I was staying, and when I told him the street, rather excitedly inquired the number. When I told him the number, which is 240, he gave a manifest start, and after one sharp glance at me sipped his water in silence for a moment. He had so evidently betrayed some ulterior knowledge with regard to my residence that I questioned him.

"'What do you know about 240 Pleasant Street?' said I.

"'Oh, nothing,' he replied, evasively, sipping his water.

"After a little while, however, he inquired, in what he evidently tried to render a casual tone, what room I occupied. 'I once lived a few weeks at 240 Pleasant Street myself,' he said. 'That house always was a boarding-house, I guess.'

"'It had stood vacant for a term of years before the present occupant rented it, I believe,' I remarked. Then I answered his question. 'I have the hall bedroom on the third floor,' said I. 'The quarters are pretty straitened, but comfortable enough as hall bedrooms go.'

"But Mr. Addison had showed such unmistakable consternation at my reply that then I persisted in my questioning as to the cause, and at last he yielded and told me what he knew. He had hesitated both because he shrank from displaying what I might consider an unmanly superstition, and because he did not wish to influence me beyond what the facts of the case warranted. 'Well, I will tell you, Wheatcroft,' he said. 'Briefly all I know is this: When last I heard of 240 Pleasant Street it was not rented because of foul play which was supposed to have taken place there, though nothing was ever proved. There were two disappearances, and—in each case—of an occupant of the hall bedroom which you now have. The first disappearance was of a very beautiful girl who had come here for her health and was said to be the victim of a profound melancholy, induced by a love disappointment. She obtained board at 240 and occupied the hall bedroom about two weeks; then one morning she was gone, having seemingly vanished into thin air. Her relatives were communicated with; she had not many, nor friends either, poor girl, and a thorough search was made, but the last I knew she had never come to light. There were two or three arrests, but nothing ever came of them. Well, that was before my day here, but the second disappearance took place when I was in the house—a fine young fellow who had overworked in college. He had to pay his own way. He had taken cold, had the grip, and that and the overwork about finished him, and he came on here for a month's rest and recuperation. He had been in that room about two weeks, a little less, when one morning he wasn't there. Then there was a great hullabaloo. It seems that he had let fall some hints to the effect that there was something queer about the room, but, of course, the police did not think much of that. They made arrests right and left, but they never found him, and the arrested were discharged, though some of them are probably under a cloud of suspicion to this day. Then the boarding-house was shut up. Six years ago nobody would have boarded there, much less occupied that hall bedroom, but now I suppose new people have come in and the story has died out. I dare say your landlady will not thank me for reviving it.'

"I assured him that it would make no possible difference to me. He looked at me sharply, and asked bluntly if I had seen anything wrong or unusual about the room. I replied, guarding myself from falsehood with a quibble, that I had seen nothing in the least unusual about the room, as indeed I had not, and have not now, but that may come. I feel that that will come in due time. Last night I neither saw, nor heard, nor smelled, nor tasted, but I—felt. Last night, having started again on my exploration of, God knows what, I had not advanced a step before I touched something. My first sensation was one of disappointment. 'It is the dresser, and I am at the end of it now,' I thought. But I soon discovered that it was not the old painted dresser which I touched, but something carved, as nearly as I could discover with my unskilled fingertips, with winged things. There were certainly long keen curves of wings which seemed to overlay an arabesque of fine leaf and flower work. I do not know what the object was that I touched. It may have been a chest. I may seem to be exaggerating when I say that it somehow failed or exceeded in some mysterious respect of being the shape of anything I had ever touched. I do not know what the material was. It was as smooth as ivory, but it did not feel like ivory; there was a singular warmth about it, as if it had stood long in hot sunlight. I continued, and I encountered other objects I am inclined to think were pieces of furniture of fashions and possibly of uses unknown to me, and about them all was the strange mystery as to shape. At last I came to what was evidently an open window of large area. I distinctly felt a soft, warm wind, yet with a crystal freshness, blow on my face. It was not the window of my hall bedroom, that I know. Looking out, I could see nothing. I only felt the wind blowing on my face.

"Then suddenly, without any warning, my groping hands to the right and left touched living beings, beings in the likeness of men and women, palpable creatures in palpable attire. I could feel the soft silken texture of their garments which swept around me, seeming to half infold me in clinging meshes like cobwebs. I was in a crowd of these people, whatever they were, and whoever they were, but, curiously enough, without seeing one of them I had a strong sense of recognition as I passed among them. Now and then a hand that I knew closed softly over mine; once an arm passed around me. Then I began to feel myself gently swept on and impelled by this softly moving throng; their floating garments seemed to fairly wind me about, and again a swift terror overcame me. I struck my match, and was back in my hall bedroom. I wonder if I had not better keep my gas burning tonight? I wonder if it be possible that this is going too far? I wonder what became of those other people, the man and the woman who occupied this room? I wonder if I had better not stop where I am?

"January 31. Last night I saw—I saw more than I can describe, more than is lawful to describe. Something which nature has rightly hidden has been revealed to me, but it is not for me to disclose too much of her secret. This much I will say, that doors and windows open into an out-of-doors to which the outdoors which we know is but a vestibule. And there is a river; there is something strange with respect to that picture. There is a river upon which one could sail away. It was flowing silently, for tonight I could only see. I saw that I was right in thinking I recognized some of the people whom I encountered the night before, though some were strange to me. It is true that the girl who disappeared from the hall bedroom was very beautiful. Everything which I saw last night was very beautiful to my one sense that could grasp it. I wonder what it would all be if all my senses together were to grasp it? I wonder if I had better not keep my gas burning tonight? I wonder—"

This finishes the journal which Mr. Wheatcroft left in his hall bedroom. The morning after the last entry he was gone. His friend, Mr. Addison, came here, and a search was made. They even tore down the wall behind the picture, and they did find something rather queer for a house that had been used for boarders, where you would think no room would have been let run to waste. They found another room, a long narrow one, the length of the hall bedroom, but narrower, hardly more than a closet. There was no window, nor door, and all there was in it was a sheet of paper covered with figures, as if somebody had been doing sums. They made a lot of talk about those figures, and they tried to make out that the fifth dimension, whatever that is, was proved, but they said afterward they didn't prove anything. They tried to make out then that somebody had murdered poor Mr. Wheatcroft and hid the body, and they arrested poor Mr. Addison, but they couldn't make out anything against him. They proved he was in the Cure all that night and couldn't have done it. They don't know what became of Mr. Wheatcroft, and now they say two more disappeared from that same room before I rented the house.

The agent came and promised to put the new room they discovered into the hall bedroom and have everything new—papered and painted. He took away the picture; folks hinted there was something queer about that, I don't know what. It looked innocent enough, and I guess he burned it up. He said if I would stay he would arrange it with the owner, who everybody says is a very queer man, so I should not have to pay much if any rent. But I told him I couldn't stay if he was to give me the rent. That I wasn't afraid of anything myself, though I must say I wouldn't want to put anybody in that hall bedroom without telling him all about it; but my boarders would leave, and I knew I couldn't get any more. I told him I would rather have had a regular ghost than what seemed to be a way of going out of the house to nowhere and never coming back again. I moved, and, as I said before, it remains to be seen whether my ill luck follows me to this house or not. Anyway, it has no hall bedroom.

—1911—

SPACE

John Buchan

John Buchan, 1st Baron Tweedsmuir (1875–1940), may seem an odd inclusion in a volume of this nature. Politician, author, lieutenant colonel—Buchan is perhaps best known for his biographies and his 1915 thriller, The Thirty-Nine Steps. *Nevertheless, he also produced several excellent strange tales. This particular story first appeared in* Blackwood's Magazine, *May 1911, before its collection in* The Moon Endureth *(1912) the following year. It is an eerie piece, told in the open air—and open space, as the story suggests, may not be as benign, or as empty, as it appears.*

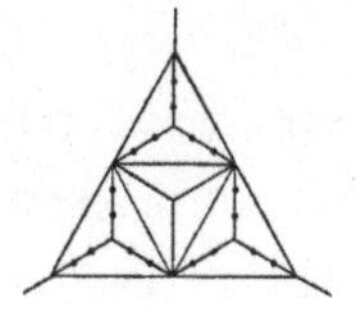

"Est impossibile? Certum est."

—TERTULLIAN

LEITHEN TOLD ME THIS STORY ONE EVENING IN EARLY SEPTEMBER as we sat beside the pony track which gropes its way from Glenvalin up the Correi na Sidhe. I had arrived that afternoon from the south, while he had been taking an off-day from a week's stalking, so we had walked up the glen together after tea to get the news of the forest. A rifle was out on the Correi na Sidhe beat, and a thin spire of smoke had risen from the top of Sgurr Dearg to show that a stag had been killed at the burn-head. The lumpish hill pony with its deer-saddle had gone up the Correi in a gillie's charge, while we followed at leisure, picking our way among the loose granite rocks and the patches of wet bogland. The track climbed high on one of the ridges of Sgurr Dearg, till it hung over a cauldron of green glen with the Alt-na-Sidhe churning in its linn a thousand feet below. It was a breathless evening, I remember, with a pale-blue sky just clearing from the haze of the day. West-wind weather may make the North, even in September, no bad imitation of the Tropics, and I sincerely pitied the man who all these stiffling hours had been toiling on the screes of Sgurr Dearg. By-and-by we sat down on a bank of heather, and idly watched the trough swimming at our feet. The clatter of the pony's hoofs grew fainter, the drone of bees had gone, even the midges seemed to have forgotten their calling. No place on earth can be so deathly still as a deer-forest early in the season before the stags have begun roaring, for there are no sheep with their homely noises, and only the rare croak of a raven breaks the silence. The hillside was far from sheer—one could have walked down with a little care—but something in the shape of the hollow and the remote gleam of white water gave it an air of extraordinary depth and space. There was a shimmer left from the day's heat, which invested bracken and rock and scree with a curious airy unreality. One could almost have believed that the eye had tricked the mind, that all was mirage, that five yards from the path the solid earth fell away into nothingness. I have a bad head, and instinctively I drew farther back into the heather. Leithen's eyes were looking vacantly before him.

"Did you ever know Hollond?" he asked.

Then he laughed shortly. "I don't know why I asked that, but somehow this place reminded me of Hollond. That glimmering hollow looks as if it were the beginning of eternity. It must be eerie to live with the feeling always on one."

Leithen seemed disinclined for further exercise. He lit a pipe and smoked quietly for a little. "Odd that you didn't know Hollond. You must have heard his name. I thought you amused yourself with metaphysics."

Then I remembered. There had been an erratic genius who had written some articles in *Mind* on that dreary subject, the mathematical conception of infinity. Men had praised them to me, but I confess I never quite understood their argument. "Wasn't he some sort of mathematical professor?" I asked.

"He was, and, in his own way, a tremendous swell. He wrote a book on Number which has translations in every European language. He is dead now, and the Royal Society founded a medal in his honour. But I wasn't thinking of that side of him."

It was the time and place for a story, for the pony would not be back for an hour. So I asked Leithen about the other side of Hollond which was recalled to him by Correi na Sidhe. He seemed a little unwilling to speak...

"I wonder if you will understand it. You ought to, of course, better than me, for you know something of philosophy. But it took me a long time to get the hang of it, and I can't give you any kind of explanation. He was my fag at Eton, and when I began to get on at the Bar I was able to advise him on one or two private matters, so that he rather fancied my legal ability. He came to me with his story because he had to tell someone, and he wouldn't trust a colleague. He said he didn't want a scientist to know, for scientists were either pledged to their own theories and wouldn't understand, or, if they understood, would get ahead of him in his researches. He wanted a lawyer, he said, who was accustomed to weighing evidence. That was good sense, for evidence must always be judged by the same laws, and I suppose in the long-run the most abstruse business comes down to a fairly simple deduction from certain data. Anyhow, that was the way he used to talk, and I listened to him, for I liked the man, and had an enormous respect for his brains. At Eton he sluiced down all the mathematics they could give him, and he was an astonishing swell at Cambridge. He was a simple fellow, too, and talked no more jargon than he could help. I used to climb with him in the Alps now and then, and you would never have guessed that he had any thoughts beyond getting up steep rocks.

"It was at Chamonix, I remember, that I first got a hint of the matter that was filling his mind. We had been talking an off-day, and were sitting in the hotel garden, watching the Aiguilles getting purple in the twilight. Chamonix always makes me choke a little—it is so crushed in by those great snow masses. I said something about it—said I liked open spaces like the Gornegrat or the Bel Alp better. He asked me why: if it was the difference of the air, or merely the wider horizon? I said it was the sense of not being crowded, of living in an empty world. He repeated the word 'empty' and laughed.

"'By "empty" you mean,' he said, 'where things don't knock up against you?'

"I told him No. I mean just empty, void, nothing but blank æther.

"You don't knock up against things here, and the air is as good as you want. It can't be the lack of ordinary emptiness you feel."

"I agreed that the word needed explaining. 'I suppose it is mental restlessness,' I said. 'I like to feel that for a tremendous distance there is nothing round me. Why, I don't know. Some men are built the other way and have a terror of space.'

"He said that that was better. 'It is a personal fancy, and depends on your *knowing* that there is nothing between you and the top of the Dent Blanche. And you know because your eyes tell you there is nothing. Even if you were blind, you might have a sort of sense about adjacent matter. Blind men often have it. But in any case, whether got from instinct or sight, the *knowledge* is what matters."

"Hollond was embarking on a Socratic dialogue in which I could see little point. I told him so, and he laughed.

"'I am not sure that I am very clear myself. But yes—there *is* a point. Supposing you knew—not by sight or by instinct, but by sheer intellectual knowledge, as I know the truth of a mathematical proposition—that what we call empty space was full, crammed. Not with lumps of what we call matter like hills and houses, but with things as real—as real to the mind. Would you still feel crowded?'

"'No,' I said, 'I don't think so. It is only what we call matter that signifies. It would be just as well not to feel crowded by the other thing, for there would be no escape from it. But what are you getting at? Do you mean atoms or electric currents or what?'

"He said he wasn't thinking about that sort of thing, and began to talk of another subject.

"Next night, when we were pigging it at the Géant *cabane*, he started again on the same tack. He asked me how I accounted for the fact that animals could find their way back over great tracts of unknown country. I said I supposed it was the homing instinct.

"'Rubbish, man,' he said. 'That's only another name for the puzzle, not an explanation. There must be some reason for it. They must *know* something that we cannot understand. Tie a cat in a bag and take it fifty miles by train and it will make its way home. That cat has some clue that we haven't.'

"I was tired and sleepy, and told him that I did not care a rush about the psychology of cats. But he was not to be snubbed, and went on talking.

"'How if Space is really full of things we cannot see and as yet do not know? How if all animals and some savages have a cell in their brain or a nerve which responds to the invisible world? How if all Space be full of these landmarks, not material in our sense, but quite real? A dog barks at nothing, a wild beast makes an aimless circuit. Why? Perhaps because Space is made up of corridors and alleys, ways to travel and things to shun? For all we know, to a greater intelligence than ours the top of Mont Blanc may be as crowded as Piccadilly Circus.'

"But at that point I fell asleep and left Hollond to repeat his questions to a guide who knew no English and a snoring porter.

"Six months later, one foggy January afternoon, Hollond rang me up at the Temple and proposed to come to see me that night after dinner. I thought he wanted to talk Alpine shop, but he turned up in Duke Street about nine with a kit-bag full of papers. He was an odd fellow to look at—a yellowish face with the skin stretched tight on the cheek-bones, clean-shaven, a sharp chin which he kept poking forward, and deep-set, greyish eyes. He was a hard fellow too, always in pretty good condition, which was remarkable considering how he slaved for nine months out of the twelve. He had a quiet, slow-spoken manner, but that night I saw that he was considerably excited.

"He said that he had come to me because we were old friends. He proposed to tell me a tremendous secret. 'I must get another mind to work on it or I'll go crazy. I don't want a scientist. I want a plain man.'

"Then he fixed me with a look like a tragic actor's. 'Do you remember that talk we had in August at Chamonix—about Space? I daresay you thought I was playing the fool. So I was in a sense, but I was feeling my way towards something which has been in my mind for ten years. Now I have got it, and you must hear about it. You may take my word that it's a pretty startling discovery.'

"I lit a pipe and told him to go ahead, warning him that I knew about as much science as the dustman.

"I am bound to say that it took me a long time to understand what he meant. He began by saying that everybody thought of Space as an 'empty homogeneous medium.' 'Never mind at present what the ultimate constituents of that medium are. We take it as a finished product, and we think of it as mere extension, something without any quality at all. That is the view of civilized man. You will find all the philosophers taking it for granted. Yes, but every living thing does not take that view. An animal, for instance. It feels a kind of quality in Space. It can find its way over new country, because it perceives certain landmarks, not necessarily material, but perceptible, or if you like intelligible. Take an Australian savage. He has the same power, and, I believe, for the same reason. He is conscious of intelligible landmarks.'

"'You mean what people call a sense of direction,' I put in.

"'Yes, but what in Heaven's name is a sense of direction? The phrase explains nothing. However incoherent the mind of the animal or the savage may be, it is there somewhere, working on some data. I've been all through the psychological and anthropological side of the business, and after you eliminate clues from sight and hearing and smell and half-conscious memory there remains a solid lump of the inexplicable.'

"Hollond's eye had kindled, and he sat doubled up in his chair, dominating me with a finger.

"'Here, then, is a power which man is civilizing himself out of. Call it anything you like, but you must admit that it is a power. Don't you see that it is a perception of another kind of reality that we are leaving behind us?... Well, you know the way nature works. The wheel comes full circle, and what we think we have lost we regain in a higher form. So for a long time I have been wondering whether the civilized mind could not recreate for itself this lost gift, the gift of seeing the quality of Space. I mean that I wondered whether the scientific modern brain could not get to the stage of realizing that Space is not an empty homogeneous medium, but full of intricate differences, intelligible and real, though not with our common reality.'

"I found all this very puzzling, and he had to repeat it several times before I got a glimpse of what he was talking about.

"'I've wondered for a long time,' he went on, 'but now, quite suddenly, I have begun to know.' He stopped and asked me abruptly if I knew much about mathematics.

"'It's a pity,' he said, 'but the main point is not technical, though I wish you could appreciate the beauty of some of my proofs.' Then he began to tell me about his last six months' work. I should have mentioned that he was a brilliant physicist besides other things. All Hollond's tastes were on the borderlands of sciences, where mathematics fades into metaphysics and physics merges in the abstrusest kind of mathematics. Well, it seems he had been working for years at the ultimate problem of matter, and especially of that rarefied matter we call æther or space. I forget what his view was—atoms or molecules or electric waves. If he ever told me I have forgotten, but I'm not certain that I ever knew. However, the point was that these ultimate constituents were dynamic and mobile, not a mere passive medium but a medium in constant movement and change. He claimed to have discovered—by ordinary inductive experiment—that the constituents of æther possessed certain functions, and moved in certain figures obedient to certain mathematical laws. Space, I gathered, was perpetually 'forming fours' in some fancy way.

"Here he left his physics and became the mathematician. Among his mathematical discoveries had been certain curves or figures or something whose behaviour involved a new dimension. I gathered that this wasn't the ordinary Fourth Dimension that people talk of, but that fourth-dimensional inwardness or involution was part of it. The explanation lay in the pile of manuscripts he left with me, but though I tried honestly I couldn't get the hang of it. My mathematics stopped with desperate finality just as he got into his subject.

"His point was that the constituents of Space moved according to these new mathematical figures of his. They were always changing, but the principles of their change were as fixed as the law of gravitation. Therefore, if you once grasped these principles you knew the contents of the void. What do you make of that?"

I said that it seemed to me a reasonable enough argument, but that it got one very little way forward. "A man," I said, "might know the contents of Space and the laws of their arrangement and yet be unable to see anything more than his fellows. It is a purely academic knowledge. His mind knows it as the result of many deductions, but his senses perceive nothing."

Leithen laughed. "Just what I said to Hollond. He asked the opinion of my legal mind. I said I could not pronounce on his argument, but that I could point out that he had established no *trait d'union* between the intellect which understood and the senses which perceived. It was like a blind man with immense knowledge but no eyes, and therefore no peg to hang his knowledge on and make it useful. He had not explained his savage or his cat. 'Hang it, man,' I said, 'before you can appreciate the existence of your Spacial forms you have to go through elaborate experiments and deductions. You can't be doing that every minute. Therefore you don't get any nearer to the *use* of the sense you say that man once possessed, though you can explain it a bit.'"

"What did he say?" I asked.

"The funny thing was that he never seemed to see my difficulty. When I kept bringing him back to it he shied off with a new wild theory of perception. He argued that the mind can live in a world of realities without any sensuous stimulus to connect them with the world of our ordinary life. Of course that wasn't my point. I supposed that this world of Space was real enough to him, but I wanted to know how he got there. He never answered me. He was the typical Cambridge man, you know—dogmatic about uncertainties, but curiously diffident about the obvious. He laboured to get me to understand the notion of his mathematical forms, which I was quite willing to take on trust from him. Some queer things he said, too. He took our feeling about Left and Right as an example of our instinct for the quality of Space. But when I objected that Left and Right varied with each object, and only existed in connection with some definite material thing, he said that that was exactly what he meant. It was an example of the mobility of the Spacial forms. Do you see any sense in that?"

I shook my head. It seemed to me pure craziness.

"And then he tried to show me what he called the 'involution of Space,' by taking two points on a piece of paper. The points were a foot away when the paper was flat, but they coincided when it was doubled up. He said that there were no gaps between the figures, for the medium was continuous, and he took as an illustration the loops on a cord. You are to think of a cord always looping and unlooping itself according to certain mathematical laws. Oh, I tell you, I gave up trying to follow him. And he was so desperately in earnest all the time. By his account Space was a sort of mathematical pandemonium."

Leithen stopped to refill his pipe, and I mused upon the ironic fate which had compelled a mathematical genius to make his sole confidant of a philistine lawyer, and induced that lawyer to repeat it confusedly to an ignoramus at twilight on a Scotch hill. As told by Leithen it was a very halting tale.

"But there was one thing I could see very clearly," Leithen went on, "and that was Hollond's own case. This crowded world of Space was perfectly real to him. How he had got to it I do not know. Perhaps his mind, dwelling constantly on the problem, had unsealed some atrophied cell and restored the old instinct. Anyhow, he was living his daily life with a foot in each world.

"He often came to see me, and after the first hectic discussions he didn't talk much. There was no noticeable change in him—a little more abstracted perhaps. He would walk in the street or come into a room with a quick look round him, and sometimes for no earthly reason he would swerve. Did you ever watch a cat crossing a room? It sidles along by the furniture and walks over an open space of carpet as if it were picking its way among obstacles. Well, Hollond behaved like that, but he had always been counted a little odd, and nobody noticed it but me.

"I knew better than to chaff him, and we had stopped argument, so there wasn't much to be said. But sometimes he would give me news about his experiences. The whole thing was perfectly clear and scientific and above-board, and nothing creepy about it. You know how I hate the washy supernatural stuff they give us nowadays. Hollond was well and fit, with an appetite like a hunter. But as he talked, sometimes—well, you know I haven't much in the way of nerves or imagination—but I used to get a little eerie. Used to feel the solid earth dissolving round me. It was the opposite of vertigo, if you understand me—a sense of airy realities crowding in on you,—crowding the mind, that is, not the body.

"I gathered from Hollond that he was always conscious of corridors and halls and alleys in Space, shifting, but shifting according to inexorable laws. I never could get quite clear as to what this consciousness was like. When I asked he used to look puzzled and worried and helpless. I made out from him that one landmark involved a sequence, and once given a bearing from an object you could keep the direction without a mistake. He told me he could easily, if he wanted, go in a dirigible from the top of Mont Blanc to the top of Snowdon in the thickest fog and without a compass, if he were given the proper angle to start from. I confess I didn't follow that myself. Material objects had nothing to do with the Spacial forms, for a table or a bed in our world might be placed across a corridor of Space. The forms played their game independent of our kind of reality. But the worst of it was, that if you kept your mind too much in one world you were apt to forget about the other, and Hollond was always barking his shins on stones and chairs and things.

"He told me all this quite simply and frankly. Remember his mind and no other part of him lived in his new world. He said it gave him an odd sense of detachment to sit in a room among people, and to know that nothing there but himself had any relation at all to the infinite strange world of Space that flowed around them. He would listen, he said, to a great man talking, with one eye on the cat on the rug, thinking to himself how much more the cat knew than the man."

"How long was it before he went mad?" I asked.

It was a foolish question, and made Leithen cross. "He never went mad in your sense. My dear fellow, you're very much wrong if you think there was anything pathological about him—then. The man was brilliantly sane. His mind was as keen as a keen sword. I couldn't understand him, but I could judge of his sanity right enough."

I asked if it made him happy or miserable.

"At first I think it made him uncomfortable. He was restless because he knew too much and too little. The unknown pressed in on his mind, as bad air weighs on the lungs. Then it lightened, and he accepted the new world in the same sober practical way that he took other things. I think that the free exercise of his mind in a pure medium gave him a feeling of extraordinary power and ease. His eyes used to sparkle when he talked. And another odd thing he told me. He was a keen rock-climber, but, curiously enough, he had never a very good head. Dizzy heights always worried him, though he managed to keep hold on himself. But now all that had gone. The sense of the fulness of Space made him as happy—happier I believe—with his legs dangling into eternity, as sitting before his own study fire.

"I remember saying that it was all rather like the mediæval wizards who made their spells by means of numbers and figures.

"He caught me up at once. 'Not numbers,' he said. 'Number has no place in Nature. It is an invention of the human mind to atone for a bad memory. But figures are a different matter. All the mysteries of the world are in them, and the old magicians knew that at least, if they knew no more.'

"He had only one grievance. He complained that it was terribly lonely. 'It is the Desolation,' he would quote, 'spoken of by Daniel the prophet.' He would spend hours travelling those eerie shifting corridors of Space with no hint of another human soul. How could there be? It was a world of pure reason, where human personality had no place. What puzzled me was why he should feel the absence of this. One wouldn't, you know, in an intricate problem of geometry or a game of chess. I asked him, but he didn't understand the question. I puzzled over it a good deal, for it seemed to me that if Hollond felt lonely, there must be more in this world of his than we imagined. I began to wonder if there was any truth in fads like psychical research. Also, I was not so sure that he was as normal as I had thought: it looked as if his nerves might be going bad.

"Oddly enough, Hollond was getting on the same track himself. He had discovered, so he said, that in sleep everybody now and then lived in this new world of his. You know how one dreams of triangular railway platforms with trains running simultaneously down all three sides and not colliding. Well, this sort of cantrip was 'common form,' as we say at the Bar, in Hollond's Space, and he was very curious about the why and wherefore of Sleep. He began to haunt psychological laboratories, where they experiment with the charwoman and the odd man, and he used to go up to Cambridge for *séances*. It was a foreign atmosphere to him, and I don't think he was very happy in it. He found so many charlatans that he used to get angry, and declare he would be better employed at Mother's Meetings!"

From far up the Glen came the sound of the pony's hoofs. The stag had been loaded up, and the gillies were returning. Leithen looked at his watch. "We'd better wait and see the beast," he said.

"... Well, nothing happened for more than a year. Then one evening in May he burst into my rooms in high excitement. You understand quite clearly that there was no suspicion of horror or fright or anything unpleasant about this world he had discovered.

It was simply a series of interesting and difficult problems. All this time Hollond had been rather extra well and cheery. But when he came in I thought I noticed a different look in his eyes, something puzzled and diffident and apprehensive.

"'There's a queer performance going on in the other world,' he said. 'It's unbelievable. I never dreamed of such a thing. I—I don't quite know how to put it, and I don't know how to explain it, but—but I am becoming aware that there are other beings—other minds—moving in Space besides mine.'

"I suppose I ought to have realized then that things were beginning to go wrong. But it was very difficult, he was so rational and anxious to make it all clear. I asked him how he knew. There could, of course, on his own showing be no *change* in that world, for the forms of Space moved and existed under inexorable laws. He said he found his own mind failing him at points. There would come over him a sense of fear—intellectual fear—and weakness, a sense of something else, quite alien to Space, thwarting him. Of course he could only describe his impressions very lamely, for they were purely of the mind, and he had no material peg to hang them on, so that I could realize them. But the gist of it was that he had been gradually becoming conscious of what he called 'Presences' in his world. They had no effect on Space—did not leave footprints in its corridors, for instance—but they affected his mind. There was some mysterious contact established between him and them. I asked him if the affection was unpleasant, and he said, 'No, not exactly.' But I could see a hint of fear in his eyes.

"Think of it. Try to realize what intellectual fear is. I can't, but it is conceivable. To you and me fear implies pain to ourselves or some other, and such pain is always in the last resort pain of the flesh.

Consider it carefully and you will see that it is so. But imagine fear so sublimated and transmuted as to be the tension of pure spirit. I can't realize it, but I think it possible. I don't pretend to understand how Hollond got to know about these Presences. But there was no doubt about the fact. He was positive, and he wasn't in the least mad—not in our sense. In that very month he published his book on Number, and gave a German professor who attacked it a most tremendous public trouncing.

"I know what you are going to say,—that the fancy was a weakening of the mind from within. I admit I should have thought of that, but he looked so confoundedly sane and able that it seemed ridiculous. He kept asking me my opinion, as a lawyer, on the facts he offered. It was the oddest case ever put before me, but I did my best for him. I dropped all my own views of sense and nonsense. I told him that, taking all that he had told me as fact, the Presences might be either ordinary minds traversing Space in sleep; or minds such as his which had independently captured the sense of Space's quality; or, finally, the spirits of just men made perfect, behaving as psychical researchers think they do. It was a ridiculous task to set a prosaic man, and I wasn't quite serious. But Hollond was serious enough.

"He admitted that all three explanations were conceivable, but he was very doubtful about the first. The projection of the spirit into Space during sleep, he thought, was a faint and feeble thing, and these were powerful Presences. With the second and the third he was rather impressed. I suppose I should have seen what was happening and tried to stop it; at least, looking back that seems to have been my duty. But it was difficult to think that anything was wrong with Hollond; indeed the odd thing is that all this time the idea of madness never entered my head. I rather backed him up. Somehow the thing took my fancy, though I thought it moonshine at the bottom of my heart. I enlarged on the pioneering before him. 'Think,' I told him, 'what may be waiting for you. You may discover the meaning of Spirit. You may open up a new world, as rich as the old one, but imperishable. You may prove to mankind their immortality and deliver them for ever from the fear of death. Why, man, you are picking at the lock of all the world's mysteries.'

"But Hollond did not cheer up. He seemed strangely languid and dispirited. 'That is all true enough,' he said, 'if you are right, if your alternatives are exhaustive. But suppose they are something else, something...' What that 'something' might be he had apparently no idea, and very soon he went away.

"He said another thing before he left. He asked me if I ever read poetry, and I said, Not often. Nor did he: but he had picked up a little book somewhere and found a man who knew about the Presences. I think his name was Traherne, one of the seventeenth-century fellows. He quoted a verse which stuck to my fly-paper memory. It ran something like this:—

"'Within the region of the air,
Compassed about with Heavens fair,
Great tracts of lands there may be found,
    Where many numerous hosts,
    In those far distant coasts,
For other great and glorious ends
Inhabit, my yet unknown friends.'

Hollond was positive he did not mean angels or anything of the sort. I told him that Traherne evidently took a cheerful view of them. He admitted that, but added: 'He had religion, you see. He believed that everything was for the best. I am not a man of faith, and can only take comfort from what I understand. I'm in the dark, I tell you...'

"Next week I was busy with the Chilian Arbitration case, and saw nobody for a couple of months. Then one evening I ran against Hollond on the Embankment, and thought him looking horribly ill. He walked back with me to my rooms, and hardly uttered one word all the way. I gave him a stiff whisky-and-soda, which he gulped down absent-mindedly. There was that strained, hunted look in his eyes that you see in a frightened animal's. He was always lean, but now he had fallen away to skin and bone.

"'I can't stay long,' he told me, 'for I'm off to the Alps tomorrow and I have a lot to do.' Before then he used to plunge readily into his story, but now he seemed shy about beginning. Indeed I had to ask him a question.

"'Things are difficult,' he said hesitatingly, 'and rather distressing. Do you know, Leithen, I think you were wrong about—about what I spoke to you of. You said there must be one of three explanations. I am beginning to think that there is a fourth...'

"He stopped for a second or two, then suddenly leaned forward and gripped my knee so fiercely that I cried out. 'That world is the Desolation,' he said in a choking voice, 'and perhaps I am getting near the Abomination of the Desolation that the old prophet spoke of. I tell you, man, I am on the edge of a terror, a terror,' he almost screamed, 'that no mortal can think of and live.'

"You can imagine that I was considerably startled. It was lightning out of a clear sky. How the devil could one associate horror with mathematics? I don't see it yet... At any rate, I— You may be sure I cursed my folly for ever pretending to take him seriously. The only way would have been to have laughed him out of it at the start. And yet I couldn't, you know—it was too real and reasonable. Anyhow, I tried a firm tone now, and told him the whole thing was arrant raving bosh. I bade him be a man and pull himself together. I made him dine with me, and took him home, and got him into a better state of mind before he went to bed. Next morning I saw him off at Charing Cross, very haggard still, but better. He promised to write to me pretty often..."

The pony, with a great eleven-pointer lurching athwart its back, was abreast of us, and from the autumn mist came the sound of soft Highland voices. Leithen and I got up to go, when we heard that the rifle had made direct for the Lodge by a short cut past the Sanctuary. In the wake of the gillies we descended the Correi road into a glen all swimming with dim purple shadows. The pony minced and boggled; the stag's antlers stood out sharp on the rise against a patch of sky, looking like a skeleton tree. Then we dropped into a covert of birches and emerged on the white glen highway.

Leithen's story had bored and puzzled me at the start, but now it had somehow gripped my fancy. Space a domain of endless corridors and Presences moving in them! The world was not quite the same as an hour ago. It was the hour, as the French say, "between dog and wolf," when the mind is disposed to marvels. I thought of my stalking on the morrow, and was miserably conscious that I would miss my stag. Those airy forms would get in the way. Confound Leithen and his yarns!

"I want to hear the end of your story," I told him, as the lights of the Lodge showed half a mile distant.

"The end was a tragedy," he said slowly. "I don't much care to talk about it. But how was I to know? I couldn't see the nerve going. You see I couldn't believe it was all nonsense. If I could I might have seen. But I still think there was something in it—up to a point. Oh, I agree he went mad in the end. It is the only explanation. Something must have snapped in that fine brain, and he saw the little bit more which we call madness. Thank God, you and I are prosaic fellows...

"I was going out to Chamonix myself a week later. But before I started I got a post-card from Hollond, the only word from him. He had printed my name and address, and on the other side had scribbled six words—*'I know at last—God's mercy.—H. G. H.'* The handwriting was like a sick man of ninety. I knew that things must be pretty bad with my friend.

"I got to Chamonix in time for his funeral. An ordinary climbing accident—you probably read about it in the papers. The Press talked about the toll which the Alps took from intellectuals—the usual rot. There was an inquiry, but the facts were quite simple. The body was only recognized by the clothes. He had fallen several thousand feet.

"It seems that he had climbed for a few days with one of the Kronigs and Dupont, and they had done some hair-raising things on the Aiguilles. Dupont told me that they had found a new route up the Montanvert side of the Charmoz. He said that Hollond climbed like a *'diable fou,'* and if you know Dupont's standard of madness you will see that the pace must have been pretty hot. 'But monsieur was sick,' he added; 'his eyes were not good. And I and

Franz, we were grieved for him and a little afraid. We were glad when he left us.'

"He dismissed the guides two days before his death. The next day he spent in the hotel, getting his affairs straight. He left everything in perfect order, but not a line to a soul, not even to his sister. The following day he set out alone about three in the morning for the Grèpon. He took the road up the Nantillons glacier to the Col, and then he must have climbed the Mummery crack by himself. After that he left the ordinary route and tried a new traverse across the Mer de Glace face. Somewhere near the top he fell, and next day a party going to the Dent du Requin found him on the rocks thousands of feet below.

"He had slipped in attempting the most foolhardy course on earth, and there was a lot of talk about the dangers of guideless climbing. But I guessed the truth, and I am sure Dupont knew, though he held his tongue..."

We were now on the gravel of the drive, and I was feeling better. The thought of dinner warmed my heart and drove out the eeriness of the twilight glen. The hour between dog and wolf was passing. After all, there was a gross and jolly earth at hand for wise men who had a mind to comfort.

Leithen, I saw, did not share my mood. He looked glum and puzzled, as if his tale had aroused grim memories. He finished it at the Lodge door.

"... For, of course, he had gone out that day to die. He had seen the something more, the little bit too much, which plucks a man from his moorings. He had gone so far into the land of pure spirit that he must needs go further and shed the fleshly envelope that cumbered him. God send that he found rest! I believe that he chose the steepest cliff in the Alps for a purpose. He wanted to be unrecognizable. He was a brave man and a good citizen. I think he hoped that those who found him might not see the look in his eyes."

—1914—

A VICTIM OF HIGHER SPACE

Algernon Blackwood

After decades of relative neglect, the literary star of Algernon Blackwood (1869–1951) is rising once again. A household name from his radio and TV work at the time of his death, his striking and sensitive fiction, filled with mystical reflection, psychological insight and a genuine sense of the unearthly has struck a chord with a new generation of readers. First published in The Occult Review, *December 1914, and collected in* Day and Night Stories, *1917, this story marks the first and only reappearance of Blackwood's famous occult detective John Silence following his introduction six years earlier in* John Silence: Physician Extraordinary *(1908). In this particular story, one mathematician's intuitive understanding of non-Euclidean geometry leads to unfortunate consequences. Perhaps the most intimate and empathetic of stories about the fourth dimension, "A Victim of Higher Space" captures the strange entanglement between mathematics and mysticism.*

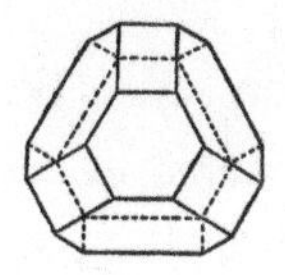

"THERE'S A HEXTRAORDINARY GENTLEMAN TO SEE YOU, sir," said the new man.

"Why 'extraordinary'?" asked Dr. Silence, drawing the tips of his thin fingers through his brown beard. His eyes twinkled pleasantly. "Why 'extraordinary,' Barker?" he repeated encouragingly, noticing the perplexed expression in the man's eyes.

"He's so—so thin, sir. I could hardly see 'im at all—at first. He was inside the house before I could ask the name," he added, remembering strict orders.

"And who brought him here?"

"He come alone, sir, in a closed cab. He pushed by me before I could say a word—making no noise not what I could hear. He seemed to move so soft like—"

The man stopped short with obvious embarrassment, as though he had already said enough to jeopardize his new situation, but trying hard to show that he remembered the instructions and warnings he had received with regard to the admission of strangers not properly accredited.

"And where is the gentleman now?" asked Dr. Silence, turning away to conceal his amusement.

"I really couldn't exactly say, sir. I left him standing in the 'all—"

The doctor looked up sharply. "But why in the hall, Barker? Why not in the waiting-room?" He fixed his piercing though kindly eyes on the man's face. "Did he frighten you?" he asked quickly.

"I think he did, sir, if I may say so. I seemed to lose sight of him as it were—" The man stammered, evidently convinced by now that he had earned his dismissal. "He come in so funny, just like a cold wind," he added boldly, setting his heels at attention and looking his master full in the face.

The doctor made an internal note of the man's halting description; he was pleased that the slight signs of psychic intuition which had induced him to engage Barker had not entirely failed at the first trial. Dr. Silence sought for this qualification in all his assistants, from secretary to serving man, and if it surrounded him with a somewhat singular crew, the drawbacks were more than compensated for on the whole by their occasional flashes of insight.

"So the gentleman made you feel queer, did he?"

"That was it, I think, sir," repeated the man stolidly.

"And he brings no kind of introduction to me—no letter or anything?" asked the doctor, with feigned surprise, as though he knew what was coming.

The man fumbled, both in mind and pockets, and finally produced an envelope.

"I beg pardon, sir," he said, greatly flustered; "the gentleman handed me this for you."

It was a note from a discerning friend, who had never yet sent him a case that was not vitally interesting from one point or another.

"Please see the bearer of this note," the brief message ran, "though I doubt if even you can do much to help him."

John Silence paused a moment, so as to gather from the mind of the writer all that lay behind the brief words of the letter. Then he looked up at his servant with a graver expression than he had yet worn.

"Go back and find this gentleman," he said, "and show him into the green study. Do not reply to his questions, or speak more than actually necessary; but think kind, helpful, sympathetic thoughts as strongly as you can, Barker. You remember what I told you about the importance of *thinking*, when I engaged you. Put curiosity out of your mind, and think gently, sympathetically, affectionately, if you can."

He smiled, and Barker, who had recovered his composure in the doctor's presence, bowed silently and went out.

There were two different reception-rooms in Dr. Silence's house. One (intended for persons who imagined they needed spiritual assistance when really they were only candidates for the asylum) had padded walls, and was well supplied with various concealed contrivances by means of which sudden violence could be instantly met and overcome. It was, however, rarely used. The other, intended for the reception of genuine cases of spiritual distress and out-of-the-way afflictions of a psychic nature, was entirely draped and furnished in a soothing deep green, calculated to induce calmness and repose of mind. And this room was the one in which Dr. Silence interviewed the majority of his "queer" cases, and the one into which he had directed Barker to show his present caller.

To begin with, the armchair in which the patient was always directed to sit, was nailed to the floor, since its immovability tended to impart this same excellent characteristic to the occupant. Patients invariably grew excited when talking about themselves, and their excitement tended to confuse their thoughts and to exaggerate their language. The inflexibility of the chair helped to counteract this. After repeated endeavours to drag it forward, or push it back, they ended by resigning themselves to sitting quietly.

And with the futility of fidgeting there followed a calmer state of mind.

Upon the floor, and at intervals in the wall immediately behind, were certain tiny green buttons, practically unnoticeable, which on being pressed permitted a soothing and persuasive narcotic to rise invisibly about the occupant of the chair. The effect upon the excitable patient was rapid, admirable, and harmless. The green study was further provided with a secret spy-hole; for John Silence liked when possible to observe his patient's face before it had assumed that mask the features of the human countenance invariably wear in the presence of another person. A man sitting alone wears a psychic expression; and this expression is the man himself. It disappears the moment another person joins him. And Dr. Silence often learned more from a few moments' secret observation of a face than from hours of conversation with its owner afterwards.

A very light, almost a dancing, step followed Barker's heavy tread towards the green room, and a moment afterwards the man came in and announced that the gentleman was waiting. He was still pale and his manner nervous.

"Never mind, Barker," the doctor said kindly; "if you were not psychic the man would have had no effect upon you at all. You only need training and development. And when you have learned to interpret these feelings and sensations better, you will feel no fear, but only a great sympathy."

"Yes, sir; thank you, sir!" And Barker bowed and made his escape, while Dr. Silence, an amused smile lurking about the corners of his mouth, made his way noiselessly down the passage and put his eye to the spy-hole in the door of the green study.

This spy-hole was so placed that it commanded a view of almost the entire room, and, looking through it, the doctor saw a hat, gloves, and umbrella lying on a chair by the table, but searched at first in vain for their owner.

The windows were both closed and a brisk fire burned in the grate. There were various signs—signs intelligible at least to a keenly intuitive soul—that the room was occupied, yet so far as human beings were concerned, it was empty, utterly empty. No one sat in the chairs; no one stood on the mat before the fire; there was no sign even that a patient was anywhere close against the wall, examining the Böcklin reproductions—as patients so often did when they thought they were alone—and therefore rather difficult to see from the spy-hole. Ordinarily speaking, there was no one in the room. It was undeniable.

Yet Dr. Silence was quite well aware that a human being *was* in the room. His psychic apparatus never failed in letting him know the proximity of an incarnate or discarnate being. Even in the dark he could tell that. And he now knew positively that his patient—the patient who had alarmed Barker, and had then tripped down the corridor with that dancing foot-step—was somewhere concealed within the four walls commanded by his spy-hole. He also realized—and this was most unusual—that this individual whom he desired to watch knew that he was being watched. And, further, that the stranger himself was also watching! In fact, that it was he, the doctor, who was being observed—and, by an observer as keen and trained as himself.

An inkling of the true state of the case began to dawn upon him, and he was on the verge of entering—indeed, his hand already touched the door-knob—when his eye, still glued to the spy-hole, detected a slight movement. Directly opposite, between him and the fireplace, something stirred. He watched very attentively and made certain that he was not mistaken. An object on the mantelpiece—it was a blue vase—disappeared from view. It passed out of sight together with the portion of the marble mantelpiece on which it rested. Next, that part of the fire and grate and brass fender immediately below it vanished entirely, as though a slice had been taken clean out of them.

Dr. Silence then understood that something between him and these objects was slowly coming into being, something that concealed them and obstructed his vision by inserting itself in the line of sight between them and himself.

He quietly awaited further results before going in.

First he saw a thin perpendicular line tracing itself from just above the height of the clock and continuing downwards till it reached the woolly fire-mat. This line grew wider, broadened, grew solid. It was no shadow; it was something substantial. It defined itself more and more. Then suddenly, at the top of the line, and about on a level with the face of the clock, he saw a round luminous disc gazing steadily at him. It was a human eye, looking straight into his own, pressed there against the spy-hole. And it was bright with intelligence. Dr. Silence held his breath for a moment—and stared back at it.

Then, like some one moving out of deep shadow into light, he saw the figure of a man come sliding sideways into view, a whitish face following the eye, and the perpendicular line he had first observed broadening out and developing into the complete figure of a human being. It was the patient. He had apparently been standing there in front of the fire all the time. A second eye had followed the first and both of them stared steadily at the spy-hole, sharply concentrated, yet with a sly twinkle of humour and amusement that made it impossible for the doctor to maintain his position any longer.

He opened the door and went in quickly. As he did so he noticed for the first time the sound of a German band coming in gaily through the open ventilators. In some intuitive, unaccountable fashion the music connected itself with the patient he was about to interview. This sort of prevision was not unfamiliar to him. It always explained itself later.

The man, he saw, was of middle age and of very ordinary appearance; so ordinary, in fact, that he was difficult to describe—his only peculiarity being his extreme thinness. Pleasant—that is, good—vibrations issued from his atmosphere and met Dr. Silence as he advanced to greet him, yet vibrations alive with currents and discharges betraying the perturbed and disordered condition of his mind and brain. There was evidently something wholly out of the usual in the state of his thoughts. Yet, though strange, it was not altogether distressing; it was not the impression that the broken and violent atmosphere of the insane produces upon the mind. Dr. Silence realized in a flash that here was a case of absorbing interest that might require all his powers to handle properly.

"I was watching you through my little peep-hole—as you saw," he began, with a pleasant smile, advancing to shake hands. "I find it of the greatest assistance sometimes—"

But the patient interrupted him at once. His voice was hurried and had odd, shrill changes in it, breaking from high to low in unexpected fashion. One moment it thundered, the next it almost squeaked.

"I understand without explanation," he broke in rapidly. "You get the true note of a man in this way—when he thinks himself unobserved. I quite agree. Only, in my case, I fear, you saw very little. My case, as you of course grasp, Dr. Silence, is extremely peculiar, uncomfortably peculiar. Indeed, unless Sir William had positively assured me—"

"My friend has sent you to me," the doctor interrupted gravely, with a gentle note of authority, "and that is quite sufficient. Pray, be seated, Mr.—"

"Mudge—Racine Mudge," returned the other.

"Take this comfortable one, Mr. Mudge," leading him to the fixed chair, "and tell me your condition in your own way and at your own pace. My whole day is at your service if you require it."

Mr. Mudge moved towards the chair in question and then hesitated.

"You will promise me not to use the narcotic buttons," he said, before sitting down. "I do not need them. Also I ought to mention that anything you think of vividly will reach my mind. That is apparently part of my peculiar case." He sat down with a sigh and arranged his thin legs and body into a position of comfort. Evidently he was very sensitive to the thoughts of others, for the picture of the green buttons had only entered the doctor's mind for a second, yet the other had instantly snapped it up. Dr. Silence noticed, too, that Mr. Mudge held on tightly with both hands to the arms of the chair.

"I'm rather glad the chair is nailed to the floor," he remarked, as he settled himself more comfortably. "It suits me admirably. The fact is—and this is my case in a nutshell—which is all that a doctor of your marvellous development requires—the fact is, Dr.

Silence, I am a victim of Higher Space. That's what's the matter with me—Higher Space!"

The two looked at each other for a space in silence, the little patient holding tightly to the arms of the chair which "suited him admirably," and looking up with staring eyes, his atmosphere positively trembling with the waves of some unknown activity; while the doctor smiled kindly and sympathetically, and put his whole person as far as possible into the mental condition of the other.

"Higher Space," repeated Mr. Mudge, "that's what it is. Now, do you think you can help me with *that*?"

There was a pause during which the men's eyes steadily searched down below the surface of their respective personalities. Then Dr. Silence spoke.

"I am quite sure I can help," he answered quietly; "sympathy must always help, and suffering always owns my sympathy. I see you have suffered cruelly. You must tell me all about your case, and when I hear the gradual steps by which you reached this strange condition, I have no doubt I can be of assistance to you."

He drew a chair up beside his interlocutor and laid a hand on his shoulder for a moment. His whole being radiated kindness, intelligence, desire to help.

"For instance," he went on, "I feel sure it was the result of no mere chance that you became familiar with the terrors of what you term Higher Space; for higher space is no mere external measurement. It is, of course, a spiritual state, a spiritual condition, an inner development, and one that we must recognize as abnormal, since it is beyond the reach of the world at the present stage of evolution. Higher space is a mystical state."

"Oh!" cried the other, rubbing his bird-like hands with pleasure, "the relief it is to me to talk to some one who can understand! Of course what you say is the utter truth. And you are right that no mere chance led me to my present condition, but, on the other hand, prolonged and deliberate study. Yet chance in a sense now governs it. I mean, my entering the condition of higher space seems to depend upon the chance of this and that circumstance. For instance, the mere sound of that German band sent me off. Not that all music will do so, but certain sounds, certain vibrations, at once key me up to the requisite pitch, and off I go. Wagner's music always does it, and that band must have been playing a stray bit of Wagner. But I'll come to all that later. Only, first, I must ask you to send away your man from the spy-hole."

John Silence looked up with a start, for Mr. Mudge's back was to the door, and there was no mirror. He saw the brown eye of Barker glued to the little circle of glass, and he crossed the room without a word and snapped down the black shutter provided for the purpose, and then heard Barker shuffle away along the passage.

"Now," continued the little man in the chair, "I can begin. You have managed to put me completely at my ease, and I feel I may tell you my whole case without shame or reserve. You will understand. But you must be patient with me if I go into details that are already familiar to you—details of higher space I mean—and if I seem stupid when I have to describe things that transcend the power of language and are really therefore indescribable."

"My dear friend," put in the other calmly, "that goes without saying. To know higher space is an experience that defies description, and one is obliged to make use of more or less intelligible symbols. But, pray, proceed. Your vivid thoughts will tell me more than your halting words."

An immense sigh of relief proceeded from the little figure half lost in the depths of the chair. Such intelligent sympathy meeting him half-way was a new experience to him, and it touched his heart at once. He leaned back, relaxing his tight hold of the arms, and began in his thin, scale-like voice.

"My mother was a Frenchwoman, and my father an Essex bargeman," he said abruptly. "Hence my name—Racine and—Mudge. My father died before I ever saw him. My mother inherited money from her Bordeaux relations, and when she died soon after, I was left alone with wealth and a strange freedom. I had no guardian, trustees, sisters, brothers, or any connections in the whole world to look after me. I grew up, therefore, utterly without education. This much was to my advantage; I learned none of that deceitful rubbish taught in schools, and so had nothing to unlearn when I awakened to my true love—mathematics, higher mathematics and higher geometry. These, however, I seemed to know instinctively. It was like the memory of what I had deeply studied before; the principles were in my blood, and I simply raced through the ordinary stages, and beyond, and then did the same with geometry. Afterwards, when I read the books on these subjects, I understood how swift and undeviating the knowledge had come back to me. It was simply memory. It was simply *re-collecting* the memories of what I had known before in a previous existence and required no books to teach me."

In his growing excitement, Mr. Mudge attempted to drag the chair forward a little nearer to his listener, and then smiled faintly as he resigned himself instantly again to its immovability, and plunged anew into the recital of his singular "disease."

"The audacious speculations of Bolyai, the amazing theories of Gauss—that through a point more than one line could be drawn parallel to a given line; the possibility that the angles of a triangle are together *greater* than two right angles, if drawn upon immense curvatures;—the breathless intuitions of Beltrami and Lobatchewsky;—all these I hurried through, and emerged, panting but unsatisfied, upon the verge of my—my new world, my higher space possibilities—in a word, my disease!

"How I got there," he resumed after a brief pause, during which he appeared to be listening intently for an approaching sound, "is more than I can put intelligibly into words. I can only hope to leave your mind with an intuitive comprehension of the possibility of what I say.

"Here, however, came a change. At this point I was no longer absorbing the fruits of studies I had made before; it was the beginning of new efforts to learn for the first time, and I had to go slowly and laboriously through terrible work. Here I sought for the theories and speculations of others. But books were few and far between, and with the exception of one man—a 'dreamer,' the world called him—whose audacity and piercing intuition amazed and delighted me beyond description, I found no one to guide or help.

"You, of course, Dr. Silence, understand something of what I am driving at with these stammering words, though you cannot perhaps yet guess what depths of pain my new knowledge brought me to, nor why an acquaintance with a new development of space should prove a source of misery and terror."

Mr. Racine Mudge, remembering that the chair would not move, did the next best thing he could in his desire to draw nearer to the attentive man facing him, and sat forward upon the very edge of the cushions, crossing his legs and gesticulating with both hands as though he saw into this region of new space he was attempting to describe, and might any moment tumble into it bodily from the edge of the chair and disappear from view. John Silence, separated from him by three paces, sat with his eyes fixed upon the thin white face opposite, noting every word and every gesture with deep attention.

"This room we now sit in, Dr. Silence, has one side open to space—to higher space. A closed box only *seems* closed. There is a way in and out of a soap bubble without breaking the skin."

"You tell me no new thing," the doctor interposed gently.

"Hence, if higher space exists and our world borders upon it and lies partially in it, it follows necessarily that we see only portions of all objects. We never see their true and complete shape. We see their three measurements, but not their fourth. The new direction is concealed from us, and when I hold this book and move my hand all round it I have not really made a complete circuit. We only perceive those portions of any object which exist in our three dimensions, the rest escapes us. But, once learn to see in higher space, and objects will appear as they actually are. Only they will thus be hardly recognizable!

"Now you may begin to grasp something of what I am coming to."

"I am beginning to understand something of what you must have suffered," observed the doctor soothingly, "for I have made similar experiments myself, and only stopped just in time—"

"You are the one man in all the world who can hear and understand, *and* sympathize," exclaimed Mr. Mudge, grasping his hand and holding it tightly while he spoke. The nailed chair prevented further excitability.

"Well," he resumed, after a moment's pause, "I procured the implements and the coloured blocks for practical experiment, and I followed the instructions carefully till I had arrived at a working conception of four-dimensional space. The tessaract, the figure whose boundaries are cubes, I knew by heart. That is to say, I knew it and saw it mentally, for my eye, of course, could never take in a new measurement, or my hands and feet handle it.

"So, at least, I thought," he added, making a wry face. "I had reached the stage, you see, when I could *imagine* in a new dimension. I was able to conceive the shape of that new figure which is intrinsically different to all we know,—the shape of the tessaract. I could perceive in four dimensions. When, therefore, I looked at a cube I could see all its sides at once. Its top was not foreshortened, nor its farther side and base invisible. I saw the whole thing out flat, so to speak. And this Tessaract was bounded by cubes! Moreover, I also saw its content—its in-sides."

"You were not yourself able to enter this new world," interrupted Dr. Silence.

"Not then. I was only able to conceive intuitively what it was like and how exactly it must look. Later, when I slipped in there and saw objects in their entirety, unlimited by the paucity of our poor three measurements, I very nearly lost my life. For, you see, space does not stop at a single new dimension, a fourth. It extends in all possible new ones, and we must conceive it as containing any number of new dimensions. In other words, there is no space at all, but only a spiritual condition. But, meanwhile, I had come to grasp the strange fact that the objects in our normal world appear to us only partially."

Mr. Mudge moved farther forward till he was balanced dangerously on the very edge of the chair. "From this starting point," he resumed, "I began my studies and experiments, and continued them for years. I had money, and I was without friends. I lived in solitude and experimented. My intellect, of course, had little part in the work, for intellectually it was all unthinkable. Never was the limitation of mere reason more plainly demonstrated. It was mystically, intuitively, spiritually that I began to advance. And what I learnt, and knew, and did, is all impossible to put into language, since it all describes experiences transcending the experiences of men. It is only some of the results—what you would call the symptoms of my disease—that I can give you, and even these must often appear absurd contradictions and impossible paradoxes.

"I can only tell you, Dr. Silence,"—his manner became exceedingly impressive,—"that I reached sometimes a point of view whence all the great puzzle of the world became plain to me, and I understood what they call in the Yoga books 'The Great Heresy of Separateness'; why all great teachers have urged the necessity of man loving his neighbour as himself; how men are all really *one*; and why the utter loss of self is necessary to salvation and the discovery of the true life of the soul."

He paused a moment and drew breath.

"Your speculations have been my own long ago," the doctor said quietly. "I fully realize the force of your words. Men are doubtless not separate at all—in the sense they imagine—"

"All this about the very much higher space I only dimly, very dimly conceived, of course," the other went on, raising his voice again by jerks; "but what did happen to me was the humbler accident of—the simpler disaster—oh dear, how shall I put it—?"

He stammered, and showed visible signs of distress.

"It was simply this," he resumed with a sudden rush of words, "that, accidentally, as the result of my years of experiment, I one day slipped bodily into the next world, the world of four dimensions, yet without knowing precisely how I got there, or how I could get back again. I discovered, that is, that my ordinary three-dimensional body was but an expression—a projection—of my higher four-dimensional body!

"Now you understand what I meant much earlier in our talk when I spoke of chance. I cannot control my entrance or exit. Certain people, certain human atmospheres, certain wandering forces, thoughts, desires even—the radiations of certain combinations of colour, and above all, the vibrations of certain kinds of music, will suddenly throw me into a state of what I can only describe as an intense and terrific inner vibration—and behold I am off! Off in the direction at right angles to all our known directions! Off in the direction the cube takes when it begins to trace the outlines of the new figure! Off into my breathless and semi-divine higher space! Off, *inside myself*, into the world of four dimensions!"

He gasped and dropped back into the depths of the immovable chair.

"And there," he whispered, his voice issuing from among the cushions, "there I have to stay until these vibrations subside, or until they do something which I cannot find words to describe properly or intelligibly to you—and then, behold, I am back again. First, that is, I disappear. Then I reappear."

"Just so," exclaimed Dr. Silence, "and that is why a few—"

"Why a few moments ago," interrupted Mr. Mudge, taking the words out of his mouth, "you found me gone, and then saw me return. The music of that wretched German band sent me off. Your intense thinking about me brought me back—when the band had stopped its Wagner. I saw you approach the peep-hole and I saw Barker's intention of doing so later. For me no interiors are hidden. I see inside. When in that state the content of your mind, as of your body, is open to me as the day. Oh dear, oh dear, oh dear!"

Mr. Mudge stopped and again mopped his brow. A light trembling ran over the surface of his small body like wind over grass. He still held tightly to the arms of the chair.

"At first," he presently resumed, "my new experiences were so vividly interesting that I felt no alarm. There was no room for it. The alarm came a little later."

"Then you actually penetrated far enough into that state to experience yourself as a normal portion of it?" asked the doctor, leaning forward, deeply interested.

Mr. Mudge nodded a perspiring face in reply.

"I did," he whispered, "undoubtedly I did. I am coming to all that. It began first at night, when I realized that sleep brought no loss of consciousness—"

"The spirit, of course, can never sleep. Only the body becomes unconscious," interposed John Silence.

"Yes, we know that—theoretically. At night, of course, the spirit is active elsewhere, and we have no memory of where and how, simply because the brain stays behind and receives no record. But I found that, while remaining conscious, I also retained memory. I had attained to the state of continuous consciousness, for at night I regularly, with the first approaches of drowsiness, entered *nolens volens* the four dimensional world.

"For a time this happened regularly, and I could not control it; though later I found a way to regulate it better. Apparently sleep is unnecessary in the higher—the four dimensional—body. Yes, perhaps. But I should infinitely have preferred dull sleep to the knowledge. For, unable to control my movements, I wandered to and fro, attracted owing to my partial development and premature arrival, to parts of this new world that alarmed me more and more. It was the awful waste and drift of a monstrous-world, so utterly different to all we know and see that I cannot even hint at the nature of the sights and objects and beings in it. More than that, I cannot even remember them. I cannot now picture them to myself even, but can recall only the *memory of the impression* they made upon me, the horror and devastating terror of it all. To be in several places at once, for instance—"

"Perfectly," interrupted John Silence, noticing the increase of the other's excitement, "I understand exactly. But now, please, tell me a little more of this alarm you experienced, and how it affected you."

"It's not the disappearing and reappearing *per se* that I mind," continued Mr. Mudge, "so much as certain other things. It's seeing people and objects in their weird entirety, in their true and complete shapes, that is so distressing. It introduces me to a world of monsters. Horses, dogs, cats, all of which I loved; people, trees, children; all that I have considered beautiful in life—everything, from a human face to a cathedral—appear to me in a different shape and aspect to all I have known before. I cannot perhaps convince you why this should be terrible, but I assure you that it is so. To hear the human voice proceeding from this novel appearance which I scarcely recognize as a human body is ghastly, simply ghastly. To see inside everything and everybody is a form of insight peculiarly distressing. To be so confused in geography as to find myself one moment at the North Pole, and the next at Clapham Junction—or possibly at both places simultaneously—is absurdly terrifying. Your imagination will readily furnish other details without my multiplying my experiences now. But you have no idea what it all means, and how I suffer."

Mr. Mudge paused in his panting account and lay back in his chair. He still held tightly to the arms as though they could keep him in the world of sanity and three measurements, and only now and again released his left hand in order to mop his face. He looked very thin and white and oddly unsubstantial, and he stared about him as though he saw into this other space he had been talking about.

John Silence, too, felt warm. He had listened to every word and had made many notes. The presence of this man had an exhilarating effect upon him. It seemed as if Mr. Racine Mudge still carried about with him something of that breathless higher-space condition he had been describing. At any rate, Dr. Silence had himself advanced sufficiently far along legitimate paths of spiritual and psychic transformations to realize that the visions of this extraordinary little person had a basis of truth for their origin.

After a pause that prolonged itself into minutes, he crossed the room and unlocked a drawer in a bookcase, taking out a small book with a red cover. It had a lock to it, and he produced a key out of his pocket and proceeded to open the covers. The bright eyes of Mr. Mudge never left him for a single second.

"It almost seems a pity," he said at length, "to cure you, Mr. Mudge. You are on the way to discovery of great things. Thought you may lose your life in the process—that is, your life here in the world of three dimensions—you would lose thereby nothing of great value—you will pardon my apparent rudeness, I know—and you might gain what is infinitely greater. Your suffering, of course, lies in the fact that you alternate between the two worlds and are never wholly in one or the other. Also, I rather imagine, though I cannot be certain of this from any personal experiments, that you have here and there penetrated even into space of more than four dimensions, and have hence experienced the terror you speak of."

The perspiring son of the Essex bargeman and the woman of Normandy bent his head several times in assent, but uttered no word in reply.

"Some strange psychic predisposition, dating no doubt from one of your former lives, has favoured the development of your 'disease'; and the fact that you had no normal training at school or college, no leading by the poor intellect into the culs-de-sac falsely called knowledge, has further caused your exceedingly rapid movement along the lines of direct inner experience. None of the knowledge you have foreshadowed has come to you through the senses, of course."

Mr. Mudge, sitting in his immovable chair, began to tremble slightly. A wind again seemed to pass over his surface and again to set it curiously in motion like a field of grass.

"You are merely talking to gain time," he said hurriedly, in a shaking voice. "This thinking aloud delays us. I see ahead what you are coming to, only please be quick, for something is going to happen. A band is again coming down the street, and if it plays—if it plays Wagner—I shall be off in a twinkling."

"Precisely. I will be quick. I was leading up to the point of how to effect your cure. The way is this: You must simply learn to *block the entrances.*"

"True, true, utterly true!" exclaimed the little man, dodging about nervously in the depths of his chair. "But how, in the name of space, is that to be done?"

"By concentration. They are all within you, these entrances, although outer causes such as colour, music and other things lead you towards them. These external things you cannot hope to destroy, but once the entrances are blocked, they will lead you only to bricked walls and closed channels. You will no longer be able to find the way."

"Quick, quick!" cried the bobbing figure in the chair. "How is this concentration to be effected?"

"This little book," continued Dr. Silence calmly, "will explain to you the way." He tapped the cover. "Let me now read out to you certain simple instructions, composed, as I see you divine, entirely from my own personal experiences in the same direction. Follow these instructions and you will no longer enter the state of higher space. The entrances will be blocked effectively."

Mr. Mudge sat bolt upright in his chair to listen, and John Silence cleared his throat and began to read slowly in a very distinct voice.

But before he had uttered a dozen words, something happened. A sound of street music entered the room through the open ventilators, for a band had begun to play in the stable mews at the back of the house—the March from *Tannhäuser*. Odd as it may seem that a German band should twice within the space of an hour enter the same mews and play Wagner, it was nevertheless the fact.

Mr. Racine Mudge heard it. He uttered a sharp, squeaking cry and twisted his arms with nervous energy round the chair. A piteous look that was not far from tears spread over his white face.

Grey shadows followed it—the grey of fear. He began to struggle convulsively.

"Hold me fast! Catch me! For God's sake, keep me here! I'm on the rush already. Oh, it's frightful!" he cried in tones of anguish, his voice as thin as a reed.

Dr. Silence made a plunge forward to seize him, but in a flash, before he could cover the space between them, Mr. Racine Mudge, screaming and struggling, seemed to shoot past him into invisibility. He disappeared like an arrow from a bow propelled at infinite speed, and his voice no longer sounded in the external air, but seemed in some curious way to make itself heard somewhere within the depths of the doctor's own being. It was almost like a faint singing cry in his head, like a voice of dream, a voice of vision and unreality.

"Alcohol, alcohol!" it cried, "give me alcohol! It's the quickest way. Alcohol, before I'm out of reach!"

The doctor, accustomed to rapid decisions and even more rapid action, remembered that a brandy flask stood upon the mantelpiece, and in less than a second he had seized it and was holding it out towards the space above the chair recently occupied by the visible Mudge. Then, before his very eyes, and long ere he could unscrew the metal stopper, he saw the contents of the closed glass phial sink and lessen as though some one were drinking violently and greedily of the liquor within.

"Thanks! Enough! It deadens the vibrations!" cried the faint voice in his interior, as he withdrew the flask and set it back upon the mantelpiece. He understood that in Mudge's present condition one side of the flask was open to space and he could drink without removing the stopper. He could hardly have had a more interesting proof of what he had been hearing described at such length.

But the next moment—the very same moment it almost seemed—the German band stopped midway in its tune—and there was Mr. Mudge back in his chair again, gasping and panting! "Quick!" he shrieked, "stop that band! Send it away! Catch hold of me! Block the entrances! Block the entrances! Give me the red book! Oh, oh, oh-h-h-h!!!"

The music had begun again. It was merely a temporary interruption. The *Tannhäuser* March started again, this time at a tremendous pace that made it sound like a rapid two-step as though the instruments played against time.

But the brief interruption gave Dr. Silence a moment in which to collect his scattering thoughts, and before the band had got through half a bar, he had flung himself forward upon the chair and held Mr. Racine Mudge, the struggling little victim of Higher Space, in a grip of iron. His arms went all round his diminutive person, taking in a good part of the chair at the same time. He was not a big man, yet he seemed to smother Mudge completely.

Yet, even as he did so, and felt the wriggling form underneath him, it began to melt and slip away like air or water. The wood of the armchair somehow disentangled itself from between his own arms and those of Mudge. The phenomenon known as the passage of matter through matter took place. The little man seemed actually to get mixed up in his own being. Dr. Silence could just see his face beneath him. It puckered and grew dark as though from some great internal effort. He heard the thin reedy voice cry in his ear to "Block the entrances, block the entrances!" and then—but how in the world describe what is indescribable?

John Silence half rose up to watch. Racine Mudge, his face distorted beyond all recognition, was making a marvellous inward movement, as though doubling back upon himself. He turned funnel-wise like water in a whirling vortex, and then appeared to break up somewhat as a reflection breaks up and divides in a distorting convex mirror. He went neither forward nor backwards, neither to the right nor the left, neither up nor down. But he went. He went utterly. He simply flashed away out of sight like a vanishing projectile.

All but one leg! Dr. Silence just had the time and the presence of mind to seize upon the left ankle and boot as it disappeared, and to this he held on for several seconds like grim death. Yet all the time he knew it was a foolish and useless thing to do.

The foot was in his grasp one moment, and the next it seemed—this was the only way he could describe it—inside his own skin and bones, and at the same time outside his hand and all round it. It seemed mixed up in some amazing way with his own flesh and blood. Then it was gone, and he was tightly grasping a draught of heated air.

"Gone! gone! gone!" cried a thick, whispering voice somewhere deep within his own consciousness. "Lost! lost! lost!" it repeated, growing fainter and fainter till at length it vanished into nothing and the last signs of Mr. Racine Mudge vanished with it.

John Silence locked his red book and replaced it in the cabinet, which he fastened with a click, and when Barker answered the bell he inquired if Mr. Mudge had left a card upon the table. It appeared that he had, and when the servant returned with it, Dr. Silence read the address and made a note of it. It was in North London.

"Mr. Mudge has gone," he said quietly to Barker, noticing his expression of alarm.

"He's not taken his 'at with him, sir."

"Mr. Mudge requires no hat where he is now," continued the doctor, stooping to poke the fire. "But he may return for it—"

"And the humbrella, sir."

"And the umbrella."

"He didn't go out *my* way, sir, if you please," stuttered the amazed servant, his curiosity overcoming his nervousness.

"Mr. Mudge has his own way of coming and going, and prefers it. If he returns by the door at any time remember to bring him instantly to me, and be kind and gentle with him and ask no questions. Also, remember, Barker, to think pleasantly, sympathetically, affectionately of him while he is away. Mr. Mudge is a very suffering gentleman."

Barker bowed and went out of the room backwards, gasping and feeling round the inside of his collar with three very hot fingers of one hand.

It was two days later when he brought in a telegram to the study. Dr. Silence opened it, and read as follows:—

"Bombay. Just slipped out again. All safe. Have blocked entrances. Thousand thanks. Address Cooks London. MUDGE."

Dr. Silence looked up and saw Barker staring at him bewilderingly. It occurred to him that somehow he knew the contents of the telegram.

"Make a parcel of Mr. Mudge's things," he said briefly, "and address them Thomas Cook & Sons, Ludgate Circus. And send them there exactly a month from today and marked 'To be called for.'"

"Yes, sir," said Barker, leaving the room with a deep sigh and a hurried glance at the waste-paper basket where his master had dropped the pink paper.

—1924—

THE PIKESTAFFE CASE

Algernon Blackwood

Our second entry for Algernon Blackwood is drawn from his 1924 collection, Tongues of Fire and Other Sketches. *This longer piece tells of a mathematician's quest to plumb the secrets of higher space from the bedroom of his lodgings. With the help of a gifted pupil, new discoveries and changes in consciousness await. Filled with humour, attention to emotional detail, and a creeping sense of uneasiness, "The Pikestaffe Case" is quintessential Blackwood. Like many of Blackwood's stories, the narrative is more fully concerned with the witness than the protagonist, in this case the landlady Miss Speke. A fourth-dimensional escapade wrapped in a missing-persons case, the story perfects the basic elements of a much earlier Blackwood tale from 1909 titled "Entrance and Exit".*

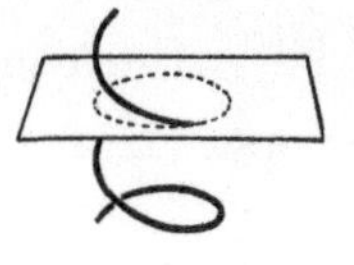

I

THE VITALITY OF OLD GOVERNESSES DESERVES AN EXPLANATORY memorandum by a good physiologist. It is remarkable. They tend to survive the grown-up married men and women they once taught as children. They hang on for ever, as a man might put it crudely, a man, that is, who, taught by one of them in his earliest schoolroom days, would answer enquiries fifty years later without enthusiasm: "Oh we keep her going, yes. She doesn't want for anything!"

Miss Helena Speke had taught the children of a distinguished family, and these distinguished children, with expensive progeny of their own now, still kept her going. They had clubbed together, seeing that Miss Speke retained her wonderful health, and had established her in a nice little house where she could take respectable lodgers—men for preference—giving them the three B's—bed, bath, and breakfast. Being a capable woman, Miss Speke more than made both ends meet. She wanted for nothing. She kept going.

Applicants for her rooms, especially for the first-floor suite, had to be recommended. She had a stern face for those who rang the bell without a letter in their pockets. She never advertised. Indeed, there was no need to do so. The two upper floors had been occupied by the same tenants for many years—a chief clerk in a branch bank and a retired clergyman respectively. It was only the best suite that sometimes "happened to be vacant at the moment." From two guineas inclusive before the war, her price for this had been raised, naturally, to four, the tenant paying his gas-stove, light, and bath extra. Breakfast—she prided herself legitimately on her good breakfasts—was included.

For a long time now this first-floor suite had been unoccupied. The cost of living worried Miss Speke, as it worried most other people. Her servant was cheap but incompetent, and once she could let the suite she meant to engage a better one. The distinguished children were scattered out of reach about the world; the eldest had been killed in the war; a married one, a woman, lived in India; another married one was in the throes of divorce—an expensive business; and the fourth, the most generous and last, found himself in the Bankruptcy Court, and so was unable to help.

It was in these conditions that Miss Speke, her vitality impaired, decided to advertise. Although she inserted the words "references essential," she meant in her heart to use her own judgment, and if a likely gentleman presented himself and agreed to pay her price, she might accept him. The clergyman and the bank official upstairs were a protection, she felt. She invariably mentioned them to applicants: "I have a clergyman of the Church of England on the top floor. He's been with me for eleven years. And a banker has the floor below. Mine is a very quiet house, you see." These words formed part of the ritual she recited in the hall, facing her proposed tenants on the linoleum by the hat-rack; and it was these words she addressed to the tall, thin, pale-faced man with scanty hair and spotless linen, who informed her that he was a tutor, a teacher of higher mathematics to the sons of various families—he mentioned some first-class names where references could be obtained—a student besides and something of an author in his leisure hours. His pupils he taught, of course, in their respective houses, one being in Belgrave Square, another in The Albany; it was only after tea, or in the evenings, that he did his own work. All this he explained briefly, but with great courtesy of manner.

Mr. Thorley was well spoken, with a gentle voice, kind, far-seeing eyes, and an air of being lonely and uncared for that touched some forgotten, dried-up spring in Miss Speke's otherwise rather cautious heart. He looked every inch a scholar—"and a gentleman," as she explained afterwards to everybody who was interested in him, these being numerous, of unexpected kinds, and all very close, not to say unpleasantly close, questioners indeed. But what chiefly influenced her in his favour was the fact, elicited in conversation, that years ago he had been a caller at the house in Portman Square where she was governess to the distinguished family. She did not exactly remember him, but he had certainly known Lady Araminta, the mother of her charges.

Thus it was that Mr. Thorley—John Laking Thorley, M.A., of Jesus College, Cambridge—was accepted by Miss Speke as tenant of her best suite on the first floor at the price mentioned, breakfast included, winning her confidence so fully that she never went to the trouble even of taking up the references he gave her. She liked him, she felt safe with him, she pitied him. He had not bargained, nor tried to beat her down. He just reflected a moment, then agreed. He proved, indeed, an exemplary lodger, early to bed and not too early to rise, of regular habits, thoughtful of the expensive new servant, careful with towels, electric light, and ink-stains, prompt in his payments, and never once troubling her with complaints or requests, as other lodgers did, not excepting the banker and the clergyman. Moreover, he was a tidy man, who never lost anything, because he invariably put everything in its proper place and thus knew exactly where to look for it. She noticed this tidiness at once.

Miss Speke, especially in the first days of his tenancy, studied him, as she studied all her lodgers. She studied his room when he was out "of a morning." At her leisure she did this, knowing he would never break in and disturb her unexpectedly. She was neither prying nor inquisitive, she assured herself, but she *was* curious. "I have a right to know something about the gentlemen who sleep under my roof with me," was the way she put it in her own mind. His clothes, she found, were ample, including evening dress, white gloves, and an opera hat. He had plenty of boots and shoes. His linen was good. His wardrobe, indeed, though a trifle uncared for, especially his socks, was a gentleman's wardrobe. Only one thing puzzled her. The full-length mirror, standing on mahogany legs—a present from the generous "child," now in the Bankruptcy Court, and, a handsome thing, a special attraction in the best suite—this fine mirror Mr. Thorley evidently did not like. The second or third morning he was with her she went to his bedroom before the servant had done it up, and saw, to her surprise, that this full-length glass stood with its back to the room. It had been placed close against the wall in a corner, its unattractive back turned outward.

"It gave me quite a shock to see it," as she said afterwards. "And such a handsome piece, too!"

Her first thought, indeed, sent a cold chill down her energetic spine. "He's cracked it!" But it was not cracked. She paused in some amazement, wondering why her new lodger had done this thing; then she turned the mirror again into its proper position, and left the room. Next morning she found it again with its face close against the wall. The following day it was the same—she turned it round, only to find it the next morning again with its back to the room.

She asked the servant, but the servant knew nothing about it.

"He likes it that way, I suppose, mum," was all Sarah said. "I never laid a 'and on it once."

Miss Speke, after much puzzled consideration, decided it must be something to do with the light. Mr. Thorley, she remembered, wore horn-rimmed spectacles for reading. She scented a mystery. It caused her a slight—oh, a very slight—feeling of discomfort. Well, if he did not like the handsome mirror, she could perhaps use it in her own room. To see it neglected hurt her a little. Not many furnished rooms could boast a full-length glass, she reflected. A few days later, meeting Mr. Thorley on the linoleum before the hat-rack, she enquired if he was quite comfortable, and if the breakfast was to his liking. He was polite and even cordial. Everything was perfect, he assured her. He had never been so well looked after. And the house was so quiet.

"And the bed, Mr. Thorley? You sleep well, I hope." She drew nearer to the subject of the mirror, but with caution. For some reason she found a difficulty in actually broaching it. It suddenly dawned upon her that there was something queer about his treatment of that full-length glass. She was by no means fanciful, Miss Speke, retired governess; only the faintest suspicion of something odd brushed her mind and vanished. But she did feel something. She found it impossible to mention the handsome thing outright.

"There's nothing you would like changed in the room, or altered?" she enquired with a smile, "or—in any way put different—perhaps?"

Mr. Thorley hesitated for a moment. A curious expression, half sad, half yearning, she thought, lit on his thoughtful face for one second and was gone. The idea of moving anything seemed distasteful to him.

"Nothing, Miss Speke, I thank you," he replied courteously, but without delay. "Everything is really *just* as I like it." Then, with a little bow, he asked: "I trust my typewriter disturbs nobody. Please let me know if it does."

Miss Speke assured him that nobody minded the typewriter in the least, nor even heard it, and, with another charming little bow and a smile, Mr. Thorley went out to give his lessons in the higher mathematics.

"There!" she reflected, "and I never even asked him!" It had been impossible.

From the window she watched him going down the street, his head bent, evidently in deep thought, his books beneath his arm, looking, she thought, every inch the gentleman and the scholar that he undoubtedly was. His personality left a very strong impression on her mind. She found herself rather wondering about him. As he turned the corner Miss Speke owned to two things that rose simultaneously in her mind: first, the relief that the lodger was out for the day and could be counted upon not to return unexpectedly; secondly, that it would interest her to slip up and see what kind of books he read. A minute later she was in his sitting-room. It was already swept and dusted, the breakfast cleared away, and the books, she saw, lay partly on the table where he had just left them and partly on the broad mantelpiece he used as a shelf. She was alone, the servant was downstairs in the kitchen. She examined Mr. Thorley's books.

The examination left her bewildered and uninspired. "I couldn't make them out at all," she put it. But they were evidently what she called costly volumes, and that she liked. "Something to do with his work, I suppose—mathematics, and all that," she decided, after turning over pages covered with some kind of hieroglyphics, symbols being a word she did not know in that connection. There was no printing, there were no sentences, there was nothing she could lay hold of, and the diagrams she thought perhaps were Euclid, or possibly astronomical. Most of the names were odd and quite unknown to her. Gauss! Minkowski! Lobachevsky! And it affronted her that some of these were German. A writer named Einstein was popular with her lodger, and that, she felt, was a pity, as well as a mistake in taste. It all alarmed her a little; or, rather she felt that touch of respect, almost of awe, pertaining to some world entirely beyond her ken. She was rather glad when the search—it was a duty—ended.

"There's nothing there," she reflected, meaning there was nothing that explained his dislike of the full-length mirror. And, disappointed, yet with a faint relief, she turned to his private papers. These, since he was a tidy man, were in a drawer. Mr. Thorley never left anything lying about. Now, a letter Miss Speke would not have thought of reading, but papers, especially learned papers, were another matter. Conscience, nevertheless, did prick her faintly as she cautiously turned over sheaf after sheaf of large white foolscap, covered with designs, and curves, and diagrams in ink, the ink he never spilt, and assuredly in his recent handwriting. And it was among these foolscap sheets that she suddenly came upon one sheet in particular that caught her attention and even startled her. In the centre, surrounded by scriggly hieroglyphics, numbers, curves and lines meaningless to her, she saw a drawing of the full-length mirror. Some of the curves ran into it and through it, emerging on the other side. She knew it was *the* mirror because its exact measurements were indicated in red ink.

This, as mentioned, startled her. What could it mean? she asked herself, staring intently at the curious sheet, as though it must somehow yield its secret to prolonged even if unintelligent enquiry. "It looks like an experiment or something," was the furthest her mind could probe into the mystery, though this, she admitted, was not very far. Holding the paper at various angles, even upside down, she examined it with puzzled curiosity, then slowly laid it down again in the exact place whence she had taken it. That faint breath of alarm had again suddenly brushed her soul, as though she approached a mystery she had better leave unsolved.

"It's very strange—" she began, carefully closing the drawer, but unable to complete the sentence even in her mind. "I don't think I like it—quite," and she turned to go out. It was just then that something touched her face, tickling one cheek, something fine as a cobweb, something in the air. She picked it away. It was a thread of silk, extremely fine, so fine, indeed, that it might almost have been a spider's web of gossamer such as one sees floating over the garden lawn on a sunny morning. Miss Speke brushed it away, giving it no further thought, and went about her usual daily duties.

II

But in her mind was established now a vague uneasiness, though so vague that at first she did not recognize it. Her thought would suddenly pause. "Now, what is it?" she would ask herself. "Something's on my mind. What is it I've forgotten?" The picture of her first-floor lodger appeared, and she knew at once. "Oh, yes, it's that mirror and the diagrams, of course." Some taut wire of alarm was quivering at the back of her mind. It was akin to those childhood alarms that pertain to the big unexplained mysteries no parent can elucidate because no parent knows. "Only God can tell that," says the parent, evading the insoluble problem. "I'd better not think about it," was the analogous conclusion reached by Miss Speke. Meanwhile the impression the new lodger's personality made upon her mind perceptibly deepened. He seemed to her full of power, above little things, a man of intense and mysterious mental life. He was constantly and somewhat possessingly in her thoughts. The mere thought of him, she found, stimulated her.

It was just before luncheon, as she returned from her morning marketing, that the servant drew her attention to certain marks upon the carpet of Mr. Thorley's sitting-room. She had discovered them as she handled the vacuum cleaner—faint, short lines drawn by dark chalk or crayons, in shape like the top or bottom right-angle of a square bracket, and sometimes with a tiny arrow shown as well. There were occasional other marks, too, that Miss Speke recognized as the hieroglyphics she called squiggles. Mistress and servant examined them together in a stooping position. They found others on the bedroom carpet, too, only these were not straight; they were small curved lines; and about the feet of the full-length mirror they clustered in a quantity, segments of circles, some large, some small. They looked as if someone had snipped off curly hair, or pared his finger-nails with sharp scissors, only considerably larger, and they were so faint that they were only visible when the sunlight fell upon them.

"I knew they was drawn on," said Sarah, puzzled, yet proud that she had found them, "because they didn't come up with the dust and fluff."

"I'll—speak to Mr. Thorley," was the only comment Miss Speke made. "I'll tell him." Her voice was not quite steady, but the girl apparently noticed nothing.

"There's all this too, please, mum." She pointed to a number of fine silk threads she had collected upon a bit of newspaper, preparatory to the dust-bin. "They was stuck on the cupboard door and the walls, stretched all across the room, but rather 'igh up. I only saw them by chance. One caught on my face."

Miss Speke stared, touched, examined for some seconds without speaking. She remembered the thread that had tickled her own cheek. She looked enquiringly round the room, and the servant, following her suggestion, indicated where the threads had been attached to walls and furniture. No marks, however, were left; there was no damage done.

"I'll mention it to Mr. Thorley," said her mistress briefly, unwilling to discuss the matter with the new servant, much less to admit that she was uncomfortably at sea. "Mr. Thorley," she added, as though there was nothing unusual, "is a high mathematician. He makes—measurements and—calculations of that sort." She had not sufficient control of her voice to be more explicit, and she went from the room aware that, unaccountably, she was trembling. She had first gathered up the threads, meaning to show them to her lodger when she demanded an explanation. But the explanation was delayed, for—to state it bluntly—she was afraid to ask him for it. She put it off till the following morning, then till the day after, and, finally, she decided to say nothing about the matter at all. "I'd better leave it, perhaps, after all," she persuaded herself. "There's no damage done, anyhow. I'd better not enquire." All the same she did not like it. By the end of the week, however, she was able to pride herself upon her restraint and tact; the marks on the carpet, rubbed out by the girl, were not renewed, and the fine threads of silk were never again found stretching through the air from wall to furniture. Mr. Thorley had evidently noticed their removal and had discontinued what he had observed was an undesirable performance. He was a scholar and a gentleman. But he was more. He was frank and straight-dealing. One morning he asked to see his landlady and told her all about it himself.

"Oh," he said in his pleasantest, easiest manner when she came into the room, "I wanted to tell you, Miss Speke—indeed, I meant to do so long before this—about the marks I made on your carpets"—he smiled apologetically—"and the silk threads I stretched. I use them for measurements—for problems I set my pupils, and one morning I left them there by mistake. The marks easily rub out. But I will use scraps of paper instead another time. I can pin these on—if you will kindly tell your excellent servant not to touch them—er—they're rather important to me." He smiled again charmingly, and his face wore the wistful, rather yearning expression that had already appealed to her. The eyes, it struck her, were very brilliant. "Any damage," he added—"though, I assure you, none is possible really—I would, of course, make good to you, Miss Speke."

"Thank you, Mr. Thorley," was all Miss Speke could find to say, so confused was her mind by troubling thoughts and questions she dared not express.

"Of course—this *is* my best suite, you see."

It was all most amicable and pleasant between them.

"I wonder—have my books come?" he asked, as he went out. "Ah, there they are, I do believe!" he exclaimed, for through the open front door a van was seen discharging a very large packing-case.

"Your books, Mr. Thorley—?" Miss Speke murmured, noting the size of the package with dismay. "But I'm afraid—you'll hardly find space to put them in," she stammered. "The rooms—er"—she did not wish to disparage them—"are so small, aren't they?"

Mr. Thorley smiled delightfully. "Oh, please do not trouble on that account," he said. "I shall find space all right, I assure you. It's merely a question of knowing where and how to put them," and he proceeded to give the men instructions.

A few days later a second case arrived.

"I'm expecting some instruments, too," he mentioned casually, "mathematical instruments," and he again assured her with his confident smile that she need have no anxiety on the score of space. Nor would he dent the walls or scrape the furniture the least little bit. There was always room, he reminded her gently again, provided one knew how to stow things away. Both books and instruments were necessary to his work. Miss Speke need feel no anxiety at all.

But Miss Speke felt more than anxiety, she felt uneasiness, she felt a singular growing dread. There lay in her a seed of distress that began to sprout rapidly. Everything arrived as Mr. Thorley had announced, case upon case was unpacked in his room by his own hands. The straw and wood she used for firing purposes, there was no mess, no litter, no untidiness, nor were walls and furniture injured in any way. What caused her dread to deepen into something bordering upon actual alarm was the fact that, on searching Mr. Thorley's rooms when he was out, she could discover no trace of any of the things that had arrived. There was no sign of either books or instruments. Where had he stored them? Where could they lie concealed? She asked herself innumerable questions, but found no answer to them. These stores, enough to choke and block the room, had been brought in through the sitting-room door. They could not possibly have been taken out again. They had *not* been taken out. Yet no trace of them was anywhere to be seen. It was very strange, she thought; indeed, it was more than strange. She felt excited. She felt a touch of hysterical alarm.

Meanwhile, thin strips of white paper, straight, angled, curved, were pinned upon the carpet; threads of finest silk again stretched overhead connecting the top of the door lintel with the window, the high cupboard with the curtain rods—yet too high to be brushed away merely by the head of anyone moving in the room. And the full-length mirror still stood with its face close against the wall.

The mystery of these aerial entanglements increased Miss Speke's alarm considerably. What could their purpose be? "Thank God," she thought, "this isn't war time!" She knew enough to realize their meaning was not "wireless." That they bore some relation to the lines on the carpet and to the diagrams and curves upon the paper, she grasped vaguely. But what it all meant baffled her and made her feel quite stupid. Where all the books and instruments had disappeared added to her bewilderment. She felt more and more perturbed. A vague, uncertain fear was worse than something definite she could face and deal with. Her fear increased. Then, suddenly, yet with a reasonable enough excuse, Sarah gave notice.

For some reason Miss Speke did not argue with the girl. She preferred to let the real meaning of her leaving remain unexpressed. She just let her go. But the fact disturbed her extraordinarily. Sarah had given every satisfaction, there had been no sign of a grievance, no complaint, the work was not hard, the pay was good. It was simply that the girl preferred to leave. Miss Speke attributed it to Mr. Thorley. She became more and more disturbed in mind. Also she found herself, more and more, avoiding her lodger, whose regular habits made such avoidance an easy matter. Knowing his hours of exit and entrance, she took care to be out of the way. At the mere sound of his step she flew to cover. The new servant, a stupid, yet not inefficient country girl, betrayed no reaction of any sort, no unfavourable reaction at any rate. Having received her instructions, Lizzie did her work without complaint from either side. She did not remove the paper and the thread, nor did she mention them. She seemed just the country clod she was. Miss Speke, however, began to have restless nights. She contracted an unpleasant habit: she lay awake—listening.

III

As the result of one of these sleepless nights she came to the abrupt conclusion that she would be happier without Mr. Thorley in the house—only she had not the courage to ask him to leave. The truth was she had not the courage to speak to him at all, much less to give him notice, however nicely.

After much cogitation she hit upon a plan that promised well: she sent him a carefully worded letter explaining that, owing to increased cost of living, she found herself compelled to raise his terms. The "raise" was more than considerable, it was unreasonable, but he paid what she demanded, sending down a cheque for three months in advance with his best compliments. The letter somehow made her tremble. It was at this stage she first became aware of the existence in her of other feelings than discomfort, uneasiness, and alarm. These other feelings, being in contradiction of her dread, were difficult to describe, but their result was plain—she did not really wish Mr. Thorley to go after all. His friendly "compliments," his refusal of her hint, caused her a secret pleasure. It was not the cheque at the increased rate that pleased her—it was simply the fact that her lodger meant to stay.

It might be supposed that some delayed sense of romance had been stirred in her, but this really was not the case at all. Her pleasure was due to another source, but to a source uncommonly obscure and very strange. She feared him, feared his presence, above all, feared going into his room, while yet there was something about the mere idea of Mr. Thorley that entranced her. Another thing may as well be told at once—she herself faced it boldly—she would enter his dreaded room, when he was out, and would deliberately linger there. There was an odd feeling in the room that gave her pleasure, and more than pleasure—happiness. Surrounded by the enigmas of his personality, by the lines and curves of white paper pinned upon her carpet, by the tangle of silken threads above her head, by the mysterious books, the more than mysterious diagrams in his drawer—yet all these, even the dark perplexity of the rejected mirror and the vanished objects, were forgotten in the curious sense of happiness she derived from merely sitting in his room. Her fear contained this other remarkable ingredient—an uncommon sense of joy, of liberty, of freedom. She felt *exaltée*.

She could not explain it, she did not attempt to do so. She would go shaking and trembling into his room, and a few minutes later this sense of uncommon happiness—of release, almost of escape, she felt it—would steal over her as though in her dried-up frozen soul spring had burst upon midwinter, as though something that crawled had suddenly most gloriously found wings. An indescribable exhilaration caught her.

Under this influence the dingy street turned somehow radiant, and the front door of her poor lodging-house opened upon blue seas, yellow sands, and mountains carpeted with flowers. Her whole life, painfully repressed and crushed down in the dull service of conventional nonentities, flashed into colour, movement, and adventure. Nothing confined her. She was no longer limited. She knew advance in all possible directions. She knew the stars. She knew escape!

An attempt has been made to describe for her what she never could have described herself.

The reaction, upon coming out again, was painful. Her life in the past as a governess, little better than a servant; her life in the present as lodging-house keeper; her struggle with servants, with taxes, with daily expenses; her knowledge that no future but a mere "living" lay in front of her until the grave was reached—these overwhelmed her with an intense depression that the contrast rendered almost insupportable. Whereas in *his* room she had perfume, freedom, liberty, and wonder—the wonder of some entirely new existence.

Thus, briefly, while Miss Speke longed for Mr. Thorley to leave her house, she became obsessed with the fear that one day he really *would* go. Her mind, it is seen, became uncommonly disturbed; her lodger's presence being undoubtedly the cause. Her nights were now more than restless, they were sleepless. Whence came, she asked herself repeatedly in the dark watches, her fear? Whence came, too, her strange enchantment?

It was at this juncture, then, that a further item of perplexity was added to her mind. Miss Speke, as has been seen, was honourably disposed; she respected the rights of others, their property as well. Yet, included in the odd mood of elation the room and its atmosphere caused her, was also a vagrant, elusive feeling that the intimate, the personal—above all, the personal—had lost their original rigidity. Small individual privacies, secrecy, no longer held their familiar meaning quite. The idea that most things in life were to be shared slipped into her. A "secret," to this expansive mood, was a childish attitude.

At any rate, it was while lingering in her lodger's attractive room one day—a habit now—that she did something that caused her surprise, yet did not shock her. She saw an open letter lying on his table—and she read it.

Rather than an actual letter, however, it seemed a note, a memorandum. It began "To J. L. T."

In a boyish writing, the meaning of the language escaped her entirely. She understood the strange words as little as she understood the phases of the moon, while yet she derived from their perusal a feeling of mysterious beauty, similar to the emotions the changes of that lovely satellite stirred in her:

"To J. L. T.

"I followed your instructions, though with intense effort and difficulty. I woke at 4 o'clock. About ten minutes later, as you said might happen, I woke a *second time*. The change into the second state was as great as the change from sleeping to waking, in the ordinary meaning of these words. But I could not remain 'awake.' I fell asleep again in about a minute—back into the usual waking state, I mean. Description in words is impossible, as you know. What I felt was too terrific to feel for long. The new energy must presently have *burned me up*. It frightened me—as you warned me it would. And this fear, no doubt, was the cause of my 'falling asleep' again so quickly.

"Cannot we arrange a Call for Help for similar occasions in future?

"G. P."

Against this note Mr. Thorley had written various strangest "squiggles"; higher mathematics, Miss Speke supposed. In the opposite margin, also in her lodger's writing, were these words:

"We must agree on a word to use when frightened. *Help*, or *Help me*, seems the best. To be uttered with the whole being."

Mr. Thorley had added a few other notes. She read them without the faintest prick of conscience. Though she understood no single sentence, a thrill of deep delight ran through her:

"It amounts, of course, to a new direction; a direction at right angles to all we know, a new direction in oneself, a new direction—in living. But it can, perhaps, be translated into mathematical terms by the intellect. This, however, only a simile at best. Cannot be experienced that way. Actual experience possible only to *changed consciousness*. But good to become mathematically accustomed to it. The mathematical experiments are worth it. They induce the mind, at any rate, to dwell upon the new direction. This helps..."

Miss Speke laid down the letter exactly where she had found it. No shame was in her. "G. P.," she knew, meant Gerald Pikestaffe; he was one of her lodger's best pupils, the one in Belgrave Square. Her feeling of mysterious elation, as already mentioned, seemed above all such matters as small secrecies or petty personal privacies. She had read a "private" letter without remorse. One feeling only caused in her a certain commonplace emotion: the feeling that, while she read the letter, her lodger was present, watching her. He seemed close behind her, looking over her shoulder almost, observing her acts, her mood, her very thoughts—yet not objecting. He was aware, at any rate, of what she did...

It was under these circumstances that she bethought herself of her old tenant, the retired clergyman on the top floor, and sought his aid. The consolation of talking to another would be something, yet when the interview began all she could manage to say was that her mind was troubled and her heart not quite as it should be, and that she "didn't know what to do about it all." For the life of her she could not find more definite words. To mention Mr. Thorley she found suddenly utterly impossible.

"Prayer," the old man interrupted her half-way, "prayer, my dear lady. Prayer, I find," he repeated smoothly, "is always the best course in all one's troubles and perplexities. Leave it to God. He knows. And in His good time He will answer." He advised her to read the Bible and Longfellow. She added Florence Barclay to the list and followed his advice. The books, however, comforted her very little.

After some hesitation she then tried her other tenant. But the "banker" stopped her even sooner than the clergyman had done. MacPherson was very prompt:

"I can give you another ten shillings or maybe half a guinea," he said briskly. "Times are deeficult, I know. But I can't do more. If that's suffeecient I shall be delighted to stay on—" and, with a nod and a quick smile that settled the matter then and there, he was through the door and down the steps on the way to his office.

It was evident that Miss Speke must face her troubles alone, a fact, for the rest, life had already taught her. The loyal, courageous spirit in her accepted the situation. The alternate moods of happiness and depression, meanwhile, began to wear her out. "If only Mr. Thorley would go! If only Mr. Thorley will not go!" For some weeks now she had successfully avoided him. He made no requests nor complaints. His habits were as regular as sunrise, his payments likewise. Not even the servant mentioned him. He became a shadow in the house.

Then, with the advent of summer-time, he came home, as it were, an hour earlier than usual. He invariably worked from 5.30 to 7.30, when he went out for his dinner. Tea he always had at a pupil's house. It was a light evening, caused by the advance of the clock, and Miss Speke, mending her underwear at the window, suddenly perceived his figure coming down the street.

She watched, fascinated. Of two instincts—to hide herself, or to wait there and catch his eye—she obeyed the latter. She had not seen him for several weeks, and a deep thrill of happiness ran through her. His walk was peculiar, she noticed at once; he did not walk in a straight line. His tall, thin outline flowed down the pavement in long, sweeping curves, yet quite steadily. He was not drunk. He came nearer; he was not twenty feet away; at ten feet she saw his face clearly, and received a shock. It was worn, and thin, and wasted, but a light of happiness, of something more than happiness indeed, shone in it. He reached the area railings. He looked up. His face seemed ablaze. Their eyes met, his with no start of recognition, hers with a steady stare of wonder. She ran into the passage, and before Mr. Thorley had time to use his latch-key she had opened the door for him herself. Little she knew, as she stood there trembling, that she stood also upon the threshold of an amazing adventure.

Face to face with him her presence of mind deserted her. She could only look up into that worn and wasted face, into those happy, severe, and brilliant eyes, where yet burned a strange expression of wistful yearning, of uncommon wonder, of something that seemed not of this world quite. Such an expression she had never seen before upon any human countenance. Its light dazzled her. There was uncommon fire in the eyes. It enthralled her. The same instant, as she stood there gazing at him without a single word, either of welcome or enquiry, it flashed across her that he needed something from her. He needed help, her help. It was a far-fetched notion, she was well aware, but it came to her irresistibly. The conviction was close to her, closer than her skin.

It was this knowledge, doubtless, that enabled her to hear without resentment the strange words he at once made use of:

"Ah, I thank you, Miss Speke, I thank you," the thin lips parting in a smile, the shining eyes lit with an emotion of more than ordinary welcome. "You cannot know what a relief it is to me to see you. You are so sound, so wholesome, so ordinary, so—forgive me, I beg—so commonplace."

He was gone past her and upstairs into his sitting-room. She heard the key turn softly. She was aware that she had not shut the front door. She did so, then went back, trembling, happy, frightened, into her own room. She had a curious, rushing feeling, both frightful and bewildering, that the room did not contain her... She was still sitting there two hours later, when she heard Mr. Thorley's step come down the stairs and leave the house. She was still sitting there when she heard him return, open the door with his key, and go up to his sitting-room. The interval might have been two minutes or two weeks, instead of two hours merely. And all this time she had the wondrous sensation that the room did not contain her. The walls and ceilings did not shut her in. She was out of the room. Escape had come very close to her. She was out of the house... out of herself as well...

IV

She went early to bed, taking this time the Bible with her. Her strange sensations had passed, they had left her gradually. She had made herself a cup of tea and had eaten a soft-boiled egg and some bread-and-butter. She felt more normal again, but her nerves were unusually sensitive. It was a comfort to know there were two men in the house with her, two worthy men, a clergyman and a banker. The Bible, the banker, the clergyman, with Mrs. Barclay and Longfellow not far from her bed, were certainly a source of comfort to her.

The traffic died away, the rumbling of the distant motor-buses ceased, and, with the passing of the hours, the night became intensely still.

It was April. Her window was opened at the top and she could smell the cool, damp air of coming spring. Soothed by the books she began to feel drowsy. She glanced at the clock—it was just on two—then blew out the candle and prepared to sleep. Her thoughts turned automatically to Mr. Thorley, lying asleep on the floor above, his threads and paper strips and mysterious diagrams all about him—when, suddenly, a voice broke through the silence with a cry for help. It was a man's voice, and it sounded a long way off. But she recognized it instantly, and she sprang out of bed without a trace of fear. It was Mr. Thorley calling, and in the voice was anguish.

"He's in trouble! In danger! He needs help! I knew it!" ran rapidly through her mind, as she lit the candle with fingers that did not tremble. The clock showed three. She had slept a full hour. She opened the door and peered into the passage, but saw no one there; the stairs, too, were empty. The call was not repeated.

"Mr. Thorley!" she cried aloud. "Mr. Thorley! Do you want anything?" And by the sound of her voice she realized how distant and muffled his own had been. "I'm coming!"

She stood there waiting, but no answer came. There was no sound. She realized the uncommon stillness of the night.

"Did you call me?" she tried again, but with less confidence. "Can I do anything for you?"

Again there was no answer; nothing stirred; the house was silent as the grave. The linoleum felt cold against her bare feet, and she stole back to get her slippers and a dressing-gown, while a hundred possibilities flashed through her mind at once. Oddly enough, she never once thought of burglars, nor of fire, nor, indeed, of any ordinary situation that required ordinary help. Why this was so she could not say. No ordinary fear, at any rate, assailed her in that moment, nor did she feel the smallest touch of nervousness about her own safety.

"Was it—I wonder—a dream?" she asked herself as she pulled the dressing-gown about her. "Did I dream that voice—?" when the thrilling cry broke forth again, startling her so that she nearly dropped the candle:

"Help! Help! Help me!"

Very distinct, yet muffled as by distance, it was beyond all question the voice of Mr. Thorley. What she had taken for anguish in it she now recognized was terror. It sounded on the floor above, it was the closed door doubtless that caused the muffled effect of distance.

Miss Speke ran along the passage instantly, and with extraordinary speed for an elderly woman; she was half-way up the stairs in a moment, when, just as she reached the first little landing by the bathroom and turned to begin the second flight, the voice came again: "Help! Help!" but this time with a difference that, truth to tell, did set her nerves unpleasantly aquiver. For there were two voices instead of one, and they were not upstairs at all. Both were below her in the passage she had just that moment left. Close they were behind her. One, moreover, was not the voice of Mr. Thorley. It was a boy's clear soprano. Both called for help together, and both held a note of terror that made her heart shake.

Under these conditions it may be forgiven to Miss Speke that she lost her balance and reeled against the wall, clutching the banisters for a moment's support. Yet her courage did not fail her. She turned instantly and quickly went downstairs again—to find the passage empty of any living figure. There was no one visible. There was only silence, a motionless hat-rack, the door of her own room slightly ajar, and shadows.

"Mr. Thorley!" she called. "Mr. Thorley!" her voice not quite so loud and confident as before. It had a whisper in it. No answer came. She repeated the words, her tone with still less volume.

Only faint echoes that seemed to linger unduly came in response. Peering into her own room she found it exactly as she had left it. The dining-room, facing it, was likewise empty. Yet a moment before she had plainly heard two voices calling for help within a few yards of where she stood. Two voices! What could it mean? She noticed now for the first time a peculiar freshness in the air, a sharpness, almost a perfume, as though all the windows were wide open and the air of coming spring was in the house.

Terror, though close, had not yet actually gripped her. That she had gone crazy occurred to her, but only to be dismissed. She was quite sane and self-possessed. The changing direction of the sounds lay beyond all explanation, but an explanation, she was positive, there must be. The odd freshness in the air was heartening, and seemed to brace her. No, terror had not yet really gripped her. Ideas of summoning the servant, the clergyman, the banker, these she equally dismissed. It was no ordinary help that was needed, not theirs at any rate. She went boldly upstairs again and knocked at Mr. Thorley's bedroom door. She knocked again and again, loud enough to waken him, if he had perchance called out in sleep, but not loud enough to disturb her other tenants. No answer came. There was no sound within. No light shone through the cracks. With his sitting-room the same conditions held.

It was the strangeness of the second voice that now stole over her with a deadly fear. She found herself cold and shivering. As she, at length, went slowly downstairs again the cries were suddenly audible once more. She heard both voices: "Help! Help! Help me!" Then silence. They were fainter this time. Far away, they sounded, withdrawn curiously into some remote distance, yet ever with the same anguish, the same terror in them as before. The direction, however, this time she could not tell at all. In a sense they seemed both close and far, both above her and below; they seemed—it was the only way she could describe the astounding thing—in any direction, or in all directions.

Miss Speke was really terrified at last. The strange, full horror of it gripped her, turning her heart suddenly to ice. The two voices, the terror in them, the extraordinary impression that they had withdrawn further into some astounding distance—this overcame her. She became appalled. Staggering into her room, she reached the bed and fell upon it in a senseless heap. She had fainted.

V

She slept late, owing probably to exhausted nerves. Though usually up and about by 7.30, it was after nine when the servant woke her. She sprawled half in the bed, half out; the candle, which luckily had extinguished itself in falling, lay upon the carpet. The events of the night came slowly back to her as she watched the servant's face. The girl was white and shaking.

"Are you ill, mum?" Lizzie asked anxiously in a whisper; then, without waiting for an answer, blurted out what she had really come in to say: "Mr. Thorley, mum! I can't get into his room. There's no answer." The girl was very frightened.

Mr. Thorley invariably had breakfast at 8 o'clock, and was out of the house punctually at 8.45.

"Was he ill in the night—perhaps—do you think?" Miss Speke said. It was the nearest she could get to asking if the girl had heard the voices. She had admirable control of herself by this time. She got up, still in her dressing-gown and slippers.

"Not that I know of, mum," was the reply.

"Come," said her mistress firmly. "We'll go in." And they went upstairs together.

The bedroom door, as the girl had said, was closed, but the sitting-room was open. Miss Speke led the way. The freshness of the night before lay still in the air, she noticed, though the windows were all closed tightly. There was an exhilarating sharpness, a delightful tang as of open space. She particularly mentions this. On the carpet, as usual, lay the strips of white paper, fastened with small pins, and the silk threads, also as usual, stretched across from lintel to cupboard, from window to bracket. Miss Speke brushed several of them from her face.

The door into the bedroom she opened, and went boldly in, followed more cautiously by the girl. "There's nothing to be afraid of," said her mistress firmly. The bed, she saw, had not been slept in. Everything was neat and tidy. The long mirror stood close against the wall, showing its ugly back as usual, while about its four feet clustered the curved strips of paper Miss Speke had grown accustomed to.

"Pull the blinds up, Lizzie," she said in a quiet voice.

The light now enabled her to see everything quite clearly. There were silken threads, she noticed distinctly, stretching from bed to window, and though both windows were closed there was this strange sweetness in the air as of a flowering spring garden. She sniffed it with a curious feeling of pleasure, of freedom, of release, though Lizzie, apparently, noticed nothing of all this.

"There's his 'at and mackintosh," the girl whispered in a frightened voice, pointing to the hooks on the door. "And the umbrella in the corner. But I don't see 'is boots, mum. They weren't put out to be cleaned."

Miss Speke turned and looked at her, voice and manner under full command. "What do you mean?" she asked.

"Mr. Thorley ain't gone out, mum," was the reply in a tremulous tone.

At that very moment a faint, distant cry was audible in a man's voice: "Help! Help!" Immediately after it a soprano, fainter still, called from what seemed even greater distance: "Help me!" The direction was not ascertainable. It seemed both in the room, yet far away outside in space above the roofs. A glance at the girl convinced Miss Speke that she had heard nothing.

"Mr. Thorley is not *here*," whispered Miss Speke, one hand upon the brass bed-rail for support.

The room was undeniably empty.

"Leave everything exactly as it is," ordered her mistress as they went out. Tears stood in her eyes, she lingered a moment on the threshold, but the sounds were not repeated. "Exactly as it is," she repeated, closing the bedroom and then the sitting-room door behind her. She locked the latter, putting the key in her pocket. Two days later, as Mr. Thorley had not returned, she informed the police. But Mr. Thorley never returned. He had disappeared completely. He left no trace. He was never heard of again, though—once—he was seen.

Yet, this is not entirely accurate perhaps, for he was seen twice, in the sense that he was seen by two persons, and though he was not "heard of," he was certainly heard. Miss Speke heard his voice from time to time. She heard it in the daytime and at night; calling for help and always with the same words she had first heard:

"Help! Help! Help me!" It sounded very far away, withdrawn into immense distance, the distance ever increasing. Occasionally she heard the boy's voice with it; they called together sometimes; she never heard the soprano voice alone. But the anguish and terror she had first noticed were no longer present. Alarm had gone out of them. It was more like an echo that she heard. Through all the hubbub, confusion and distressing annoyance of the police search and enquiry, the voice and voices came to her, though she never mentioned them to a single living soul, not even to her old tenants, the clergyman and the banker. They kept their rooms on—which was about all she could have asked of them. The best suite was never let again. It was kept locked and empty. The dust accumulated. The mirror remained untouched, its face against the wall.

The voices, meanwhile, grew more and more faint; the distance seemed to increase; soon the voice of the boy was no longer heard at all, only the cry of Mr. Thorley, her mysterious but perfect lodger, sang distantly from time to time, both in the sunshine and in the still darkness of the night hours. The direction whence it came, too, remained, as before, undeterminable. It came from anywhere and everywhere—from above, below, on all sides. It had become, too, a pleasant, even a happy sound; no dread belonged to it any more. The intervals grew longer then; days first, then weeks passed without a sound; and invariably, after these increasing intervals, the voice had become fainter, weaker, withdrawn into ever greater and greater distance. With the coming of the warm spring days it grew almost inaudible. Finally, with the great summer heats, it died away completely.

VI

The disappearance of Mr. Thorley, however, had caused no public disturbance on its own account, nor until it was bracketed with another disappearance, that of one of his pupils, Sir Mark Pikestaffe's son. The Pikestaffe Case then became a daily mystery that filled the papers. Mr. Thorley was of no consequence, whereas Sir Mark was a figure in the public eye.

Mr. Thorley's life, as enquiry proved, held no mystery. He had left everything in order. He did not owe a penny. He owned, indeed, considerable property, both in land and securities, and teaching mathematics, especially to promising pupils, seemed to have been a hobby merely. A half-brother called eventually to take away his few possessions, but the books and instruments he had brought into the lodging-house were never traced. He was a scholar and a gentleman to the last, a man, too, it appeared, of immense attainments and uncommon ability, one of the greatest mathematical brains, if the modest obituaries were to be believed, the world has ever known. His name now passed into oblivion. He left no record of his researches or achievements. Out of some mysterious sense of loyalty and protection Miss Speke never mentioned his peculiar personal habits. The strips of paper, as the silken threads, she had carefully removed and destroyed long before the police came to make their search of his rooms...

But the disappearance of young Gerald Pikestaffe raised a tremendous hubbub. It was some days before the two disappearances were connected, both having occurred on the same night, it was then proved. The boy, a lad of great talent, promising a brilliant future, and the favourite pupil of the older man, his tutor, had not even left the house. His room was empty—and that was all. He left no clue, no trace. Terrible hints and suggestions were, of course, spread far and wide, but there was not a scrap of evidence forthcoming to support them. Gerald Pikestaffe and Mr. Thorley, at the same moment of the same night, vanished from the face of the earth and were no more seen. The matter ended there. The one link between them appeared to have been an amazing, an exceptional gift for higher mathematics. The Pikestaffe Case merely added one more to the insoluble mysteries with which commonplace daily life is sprinkled.

It was some six weeks to a month after the event that Miss Speke received a letter from one of her former charges, the most generous one, now satisfactorily finished with the Bankruptcy Court. He had honourably discharged his obligations; he was doing well; he wrote and asked Miss Speke to put him up for a week or two. "And do *please* give me Mr. Thorley's room," he asked. "The case thrilled me, and I should like to sleep in that room. I always loved mysteries, you remember... There's something *very* mysterious about this thing. Besides, I knew the P. boy a little—an astounding genius, if ever there was one."

Though it cost her much effort and still more hesitation, she consented finally. She prepared the rooms herself. There was a new servant, Lizzie having given notice the day after the disappearance, and the older woman who now waited upon the clergyman and the banker was not quite to be trusted with the delicate job. Miss Speke, entering the empty rooms on tiptoe, a strange trepidation in her heart, but that same heart firm with courage, drew up the blinds, swept the floors, dusted the furniture, and made the bed. All she did with her own hands. Only the full-length mirror she did not touch. What terror still was in her clung to that handsome piece. It was haunted by memories. For her it was still both wonderful and somehow awful. The ghost of her strange experience hid invisibly in its polished, if now unseen, depths. She dared not handle it, far less move it from the resting-place where it rested in peace. *His* hands had placed it there. To her it was sacred.

It had been given to her by Colonel Lyle, who would now occupy the room, stand on the wondrous carpet, move through the air where once the mysterious silks had floated, sleep in the very bed itself. All this he could do, but the mirror he must not touch.

"I'll explain to him a little. I'll beg him not to move it. He's very understanding," she said to herself, as she went out to buy some flowers for the sitting-room. Colonel Lyle was expected that very afternoon. Lilac, she remembered, was what he always liked. It took her longer than she expected to find really fresh bunches, of the colour that he preferred, and when she got back it was time to be thinking about his tea. The sun's rays fell slanting down the dingy street, touching it with happy gold. This, with thoughts of the tea-kettle and what vase would suit the flowers best, filled her mind as she passed along the linoleum in the narrow hall—then noticed suddenly a new hat and coat hanging on the usually empty pegs. Colonel Lyle had arrived before his time.

"He's already come," she said to herself with a little gasp. A heavy dread settled instantly on her spirit. She stood a moment motionless in the passage, the lilac blossoms in her hand. She was listening.

"The gentleman's come, mum," she heard the servant say, and at the same moment saw her at the top of the kitchen stairs in the hall. "He went up to his room, mum."

Miss Speke held out the flowers. With an effort to make her voice sound ordinary she gave an order about them. "Put them in water, Mary, please. The double vase will do." She watched the woman take them slowly, oh, so slowly, from her. But her mind was elsewhere. It was still listening. And after the woman had gone down to the kitchen again slowly, oh, so slowly, she stood motionless for some minutes, listening, still intently listening. But no sound broke the quiet of the afternoon. She heard only the blundering noises made by the woman in the kitchen below. On the floor above was—silence.

Miss Speke then turned and went upstairs.

Now, Miss Speke admits frankly that she was "in a state," meaning thereby, doubtless, that her nerves were tightly strung. Her heart was thumping, her ears and eyes strained to their utmost capacity; her hands, she remembers, felt a little cold, and her legs moved uncertainly. She denies, however, that her "state," though it may be described as nervous, could have betrayed her into either invention or delusion. What she saw she saw, and nothing can shake her conviction. Colonel Lyle, besides, is there to support her in the main outline, and Colonel Lyle, when first he had entered the room, was certainly not "in a state," whatever excuses he may have offered later to comfort her. Moreover, to counteract her trepidation, she says that, as she pushed the door wide open—it was already ajar—the original mood of elation met her in the face with its lift of wonder and release. This modified her dread. She declares that joy rushed upon her, and that her "nerves" were on the instant entirely forgotten.

"What I saw I saw," remains her emphatic and unshakable verdict. "I saw—everything."

The first thing she saw admitted certainly of no doubt. Colonel Lyle lay huddled up against the further wall, half upon the carpet and half-leaning on the wainscoting. He was unconscious. One arm was stretched towards the mirror, the hand still clutching one of its mahogany feet. And the mirror had been moved. It turned now slightly more towards the room.

The picture, indeed, told its own story, a story Colonel Lyle himself repeated afterwards when he had recovered. He was surprised to find the mirror—his mirror—with its face to the wall; he went forward to put it in its proper position; in doing this he looked into it; he saw something, and—the next thing he knew—Miss Speke was bringing him round.

She explains, further, that her overmastering curiosity to look into the mirror, as Colonel Lyle had evidently looked himself, prevented her from immediately rendering first-aid to that gentleman, as she unquestionably should have done. Instead, she crossed the room, stepped over his huddled form, turned the mirror a little further round towards her, and looked straight into it.

The eye, apparently, takes in a great deal more than the mind is consciously aware of having "seen" at the moment. Miss Speke saw everything, she claims. But details certainly came back to her later, details she had not been aware of at the time. At the moment, however, her impressions, though extremely vivid, were limited to certain outstanding items. These items were—that her own reflection was not visible, no picture of herself being there; that Mr. Thorley and a boy—she recognized the Pikestaffe lad from the newspaper photographs she had seen—were plainly there, and that books and instruments in great quantity filled all the nearer space, blocking up the foreground. Beyond, behind, stretching in all directions, she affirms, was empty space that produced upon her the effect of the infinite heavens as seen in a clear night sky. This space was prodigious, yet in some way not alarming. It did not terrify; rather it comforted, and, in a sense, uplifted. A diffused soft light pervaded the huge panorama. There were no shadows, there were no high lights.

Curiously enough, however, the absence of any reproduction of herself did not at first strike her as at all out of the way; she noticed the fact, no more than that; it was, perhaps, naturally, the deep shock of seeing Mr. Thorley and the boy that held her absolutely spell-bound, arresting her faculties as though they had been frozen.

Mr. Thorley was moving to and fro, his body bent, his hand thrown forward. He looked as natural as in life. He moved steadily, as with a purpose, now nearer, now further, but his figure always bent as though he were intent upon something in his hands. The boy moved, too, but with a more gentle, less vigorous, motion that suggested floating. He followed the larger figure, keeping close, his face raised from time to time as though his companion spoke to him. The expression that he wore was quiet, peaceful, happy, and intent. He was absorbed in what he was doing at the moment. Then, suddenly, Mr. Thorley straightened himself up. He turned. Miss Speke saw his face for the first time. He looked into her eyes. The face blazed with light. The gaze was straight, and full, and clear. It betrayed recognition. Mr. Thorley smiled at her.

In a very few seconds she was aware of all this, of its main outlines, at any rate. She saw the moving, living figures in the midst of this stupendous and amazing space. The overwhelming surprise it caused her prevented, apparently, the lesser emotion of personal alarm; fear she certainly did not feel at first. It was when

Mr. Thorley looked at her with his brilliant eyes and blazing smile that her heart gave its violent jump, missed a beat or two, then began hammering against her ribs like released machinery that has gone beyond control. She was aware of the happy glory in the face, a face that was thin to emaciation, almost transparent, yet wearing an expression that was no longer earthly. Then, as he smiled, he came towards her; he beckoned; he stretched both hands out, while the boy looked up and watched.

Mr. Thorley's advance, however, had two distracting peculiarities—that as he drew nearer he moved not in a straight line, but in a curve. As a skater performing "edges," though on both feet instead of on one, he swept gracefully and with incredible speed in her direction. The other peculiarity was that with each step nearer his figure grew smaller. It lessened in height. He seemed, indeed, to be moving in two directions at once. He became diminutive.

The sight ought by rights to have paralysed her, yet it produced again, instead of terror, an effect of exhilaration she could not possibly account for. There came once again that fine elation to her mind. Not only did all desire to resist die away almost before it was born, but more, she felt its opposite—an overpowering wish to join him. The tiny hands were still stretched out to greet her, to draw her in, to welcome her; the smile upon the diminutive face, as it came nearer and nearer, was enchanting. She heard his voice then:

"Come, come to us! Here reality is nearer, and there is liberty...!"

The voice was very close and loud as in life, but it was not in front. It was behind her. Against her very ear it sounded in the air behind her back. She moved one foot forward; she raised her arms. She felt herself being sucked in—into that glorious space. There was an indescribable change in her whole being.

The cumulative effect of so many amazing happenings, all of them contrary to nature, should have been destructive to her reason. Their combined shock should have dislocated her system somewhere and have laid her low. But with every individual, it seems, the breaking-point is different. Her system, indeed, was dislocated, and a moment later and she was certainly laid low, yet it was not the effect of the figure, the voice, the gliding approach of Mr. Thorley that produced this. It was the flaw of little human egoism that brought her down. For it was in this instant that she first *realized* the absence of her own reflection in the mirror. The fact, though noticed before, had not entered her consciousness as such. It now definitely did so. The arms she lifted in greeting had no reflected counterpart. Her figure, she realized with a shock of terror, was not there. She dropped, then, like a stricken animal, one outstretched hand clutching the frame of the mirror as she did so.

"Gracious God!" she heard herself scream as she collapsed. She heard, too, the crash of the falling mirror which she overturned and brought down with her.

Whether the noise brought Colonel Lyle round, or whether it was the combined weight of Miss Speke and the handsome piece upon his legs that roused him, is of no consequence. He stirred, opened his eyes, disentangled himself and proceeded, not without astonishment, to render first-aid to the unconscious lady.

The explanations that followed are, equally, of little consequence. His own attack, he considered, was chiefly due to fatigue, to violent indigestion, and to the after-effects of his protracted bankruptcy proceedings. Thus, at any rate, he assured Miss Speke. He added, however, that he had received rather a shock from the handsome piece, for, surprised at finding it turned to the wall, he had replaced it and looked into it, but had not seen himself reflected. This had amazed him a good deal, yet what amazed him still more was that he had seen something moving in the depths of the glass. "I saw a face," he said, "and it was a face I knew. It was Gerald Pikestaffe. Behind him was another figure, the figure of a man, whose face I could not see." A mist rose before his eyes, his head swam a bit, and he evidently swayed for some unaccountable reason. It was a blow received in falling that stunned him momentarily.

He stood over her, while he fanned her face; her swoon was of brief duration; she recovered quickly; she listened to his story with a quiet mind. The after-effect of too great wonder leaves no room for pettier emotions, and traces of the exhilaration she had experienced were still about her heart and soul.

"Is it smashed?" was the first thing she asked, to which Colonel Lyle made no answer at first, merely pointing to the carpet where the frame of the long mirror lay in broken fragments.

"There was no glass, you see," he said presently. He, too, was quiet, his manner very earnest; his voice, though subdued as by a hint of awe, betrayed the glow of some intense inner excitement that lit fire in his eyes as well. "He had cut it out long ago, of course. He used the empty framework merely."

"Eh?" said Miss Speke, looking down incredulously, but finding no sign of splinters on the floor.

Her companion smiled. "We shall find it about somewhere if we look," he said calmly, which, indeed, proved later true—lying flat beneath the carpet under the bed. "His measurements and calculations led—probably by chance—towards the mirror"—he seemed speaking to himself more than to his bewildered listener—"perhaps by chance, perhaps by knowledge," he continued, "up to the mirror—and then *through* it." He looked down at Miss Speke and laughed a little. "So, like Alice, he went through it, too, taking his books and instruments, the boy as well, all with him. The boy, that is, had the knowledge too."

"I only know one thing," said Miss Speke, unable to follow him or find meaning in his words, "I shall never let these rooms again. I shall lock them up."

Her companion collected the broken pieces and made a little heap of them.

"And I shall pray for him," added Miss Speke, as he led her presently downstairs to her own quarters. "I shall never cease to pray for him as long as I live."

"He hardly needs that," murmured Colonel Lyle, but to himself. "The first terror has long since left him. He's found the new direction—and moved along it."

—1929—

THE HOUNDS OF TINDALOS

Frank Belknap Long

In a career spanning seventy years, Frank Belknap Long (1901–1994) produced an immense body of work amounting to hundreds of short stories, poems, and articles, and some twenty-nine novels. A close friend and correspondent of H. P. Lovecraft, this story is acknowledged to be one of the first "Cthulhu Mythos" stories not penned by Lovecraft himself. Appearing first in Weird Tales, *March 1929, before its collection seventeen years later in* The Hounds of Tindalos *(1946), Long's story invokes Theosophy, Einstein, medieval alchemy and Taoism as a backdrop for a truly terrifying encounter with entities from the fourth dimension.*

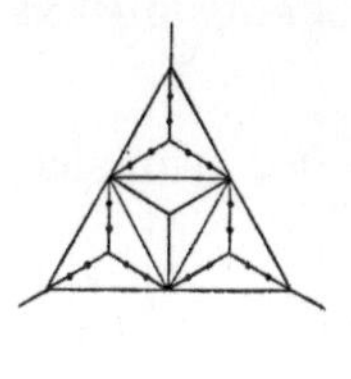

I

"I'M GLAD YOU CAME," SAID CHALMERS. HE WAS SITTING BY the window and his face was very pale. Two tall candles guttered at his elbow and cast a sickly amber light over his long nose and slightly receding chin. Chalmers would have nothing modern about his apartment. He had the soul of a mediæval ascetic, and he preferred illuminated manuscripts to automobiles and leering stone gargoyles to radios and adding-machines.

As I crossed the room to the settee he had cleared for me I glanced at his desk and was surprised to discover that he had been studying the mathematical formulæ of a celebrated contemporary physicist, and that he had covered many sheets of thin yellow paper with curious geometric designs.

"Einstein and John Dee are strange bedfellows," I said as my gaze wandered from his mathematical charts to the sixty or seventy quaint books that comprised his strange little library. Plotinus and Emanuel Moscopulus, St. Thomas Aquinas and Frenicle de Bessy stood elbow to elbow in the sombre ebony bookcase, and chairs, table and desk were littered with pamphlets about mediæval sorcery and witchcraft and black magic, and all of the valiant glamorous things that the modern world has repudiated.

Chalmers smiled engagingly, and passed me a Russian cigarette on a curiously carved tray. "We are just discovering now," he said, "that the old alchemists and sorcerers were two-thirds *right*, and that your modern biologist and materialist is nine-tenths *wrong*."

"You have always scoffed at modern science," I said, a little impatiently.

"Only at scientific dogmatism," he replied. "I have always been a rebel, a champion of originality and lost causes; that is why I have chosen to repudiate the conclusions of contemporary biologists."

"And Einstein?" I asked.

"A priest of transcendental mathematics!" he murmured reverently. "A profound mystic and explorer of the great *suspected.*"

"Then you do not entirely despise science."

"Of course not," he affirmed. "I merely distrust the scientific positivism of the past fifty years, the positivism of Haeckel and Darwin and of Mr. Bertrand Russell. I believe that biology has failed pitifully to explain the mystery of man's origin and destiny."

"Give them time," I retorted.

Chalmers' eyes glowed. "My friend," he murmured, "your pun is sublime. Give them *time*. That is precisely what I would do. But your modern biologist scoffs at time. He has the key but he refuses to use it. What do we know of time, really? Einstein believes that it is relative, that it can be interpreted in terms of space, of *curved* space. But must we stop there? When mathematics fails us can we not advance by—insight?"

"You are treading on dangerous ground," I replied. "That is a pitfall that your true investigator avoids. That is why modern science has advanced so slowly. It accepts nothing that it can not demonstrate. But you—"

"I would take hashish, opium, all manner of drugs. I would emulate the sages of the East. And then perhaps I would apprehend—"

"What?"

"The fourth dimension."

"Theosophical rubbish!"

"Perhaps. But I believe that drugs expand human consciousness. William James agreed with me. And I have discovered a new one."

"A new drug?"

"It was used centuries ago by Chinese alchemists, but it is virtually unknown in the West. Its occult properties are amazing. With its aid and the aid of my mathematical knowledge I believe that I can *go back through time*."

"I do not understand."

"Time is merely our imperfect perception of a new dimension of space. Time and motion are both illusions. Everything that has existed from the beginning of the world *exists now*. Events that occurred centuries ago on this planet continue to exist in another dimension of space. Events that will occur centuries from now *exist already*. We can not perceive their existence because we can not enter the dimension of space that contains them. Human beings as we know them are merely fractions, infinitesimally small fractions of one enormous whole. Every human being is linked with *all* the life that has preceded him on this planet. All of his ancestors are parts of him. Only time separates him from his forebears, and time is an illusion and does not exist."

"I think I understand," I murmured.

"It will be sufficient for my purpose if you can form a vague idea of what I wish to achieve. I wish to strip from my eyes the veils of illusion that time has thrown over them, and see the *beginning and the end*."

"And you think this new drug will help you?"

"I am sure that it will. And I want you to help me. I intend to take the drug immediately. I can not wait. I must *see*." His eyes glittered strangely. "I am going back, back through time."

He rose and strode to the mantel. When he faced me again he was holding a small square box in the palm of his hand. "I have here five pellets of the drug Liao. It was used by the Chinese philosopher Lao Tze, and while under its influence he visioned Tao. Tao is the most mysterious force in the world; it surrounds and pervades all things; it contains the visible universe and everything that we call reality. He who apprehends the mysteries of Tao sees clearly all that was and will be."

"Rubbish!" I retorted.

"Tao resembles a great animal, recumbent, motionless, containing in its enormous body all the worlds of our universe, the past, the present and the future. We see portions of this great monster through a slit, which we call time. With the aid of this drug I shall enlarge the slit. I shall behold the great figure of life, the great recumbent beast in its entirety."

"And what do you wish me to do?"

"Watch, my friend. Watch and take notes. And if I go back too far you must recall me to reality. You can recall me by shaking me violently. If I appear to be suffering acute physical pain you must recall me at once."

"Chalmers," I said, "I wish you wouldn't make this experiment. You are taking dreadful risks. I don't believe that there is any fourth dimension and I emphatically do not believe in Tao. And I don't approve of your experimenting with unknown drugs."

"I know the properties of this drug," he replied. "I know precisely how it affects the human animal and I know its dangers. The risk does not reside in the drug itself. My only fear is that I may become lost in time. You see, I shall assist the drug. Before I swallow this pellet I shall give my undivided attention to the geometric and algebraic symbols that I have traced on this paper." He raised the mathematical chart that rested on his knee. "I shall prepare my mind for an excursion into time. I shall *approach* the fourth dimension with my conscious mind before I take the drug which will enable me to exercise occult powers of perception. Before I enter the dream world of the Eastern mystics I shall acquire all of the mathematical help that modern science can offer. This mathematical knowledge, this conscious approach to an actual apprehension of the fourth dimension of time will supplement the work of the drug. The drug will open up stupendous new vistas—the mathematical preparation will enable me to grasp them intellectually. I have often grasped the fourth dimension in dreams, emotionally, intuitively, but I have never been able to recall, in waking life, the occult splendours that were momentarily revealed to me.

"But with your aid, I believe that I can recall them. You will take down everything that I say while I am under the influence of the drug. No matter how strange or incoherent my speech may become you will omit nothing. When I awake I may be able to supply the key to whatever is mysterious or incredible. I am not sure that I shall succeed, but if I *do* succeed"—his eyes were strangely luminous—*"time will exist for me no longer!"*

He sat down abruptly. "I shall make the experiment at once. Please stand over there by the window and watch. Have you a fountain pen?"

I nodded gloomily and removed a pale green Waterman from my upper vest pocket.

"And a pad, Frank?"

I groaned and produced a memorandum book. "I emphatically disapprove of this experiment," I muttered. "You're taking a frightful risk."

"Don't be an asinine old woman!" he admonished. "Nothing that you can say will induce me to stop now. I entreat you to remain silent while I study these charts."

He raised the charts and studied them intently. I watched the clock on the mantel as it ticked out the seconds, and a curious dread clutched at my heart so that I choked.

Suddenly the clock stopped ticking, and exactly at that moment Chalmers swallowed the drug.

I rose quickly and moved toward him, but his eyes implored me not to interfere. "The clock has stopped," he murmured. "The forces that control it approve of my experiment. *Time* stopped, and I swallowed the drug. I pray God that I shall not lose my way."

He closed his eyes and leaned back on the sofa. All of the blood had left his face and he was breathing heavily. It was clear that the drug was acting with extraordinary rapidity.

"It is beginning to get dark," he murmured. "Write that. It is beginning to get dark and the familiar objects in the room are fading out. I can discern them vaguely through my eyelids but they are fading swiftly."

I shook my pen to make the ink come and wrote rapidly in shorthand as he continued to dictate.

"I am leaving the room. The walls are vanishing and I can no longer see any of the familiar objects. Your face, though, is still visible to me. I hope that you are writing. I think that I am about to make a great leap—a leap through space. Or perhaps it is through time that I shall make the leap. I can not tell. Everything is dark, indistinct."

He sat for a while silent, with his head sunk upon his breast. Then suddenly he stiffened and his eyelids fluttered open. "God in heaven!" he cried. "I *see!*"

He was straining forward in his chair, staring at the opposite wall. But I knew that he was looking beyond the wall and that the objects in the room no longer existed for him. "Chalmers," I cried, "Chalmers, shall I wake you?"

"Do not!" he shrieked. "I see *everything*. All of the billions of lives that preceded me on this planet are before me at this moment. I see men of all ages, all races, all colours. They are fighting, killing, building, dancing, singing. They are sitting about rude fires on lonely grey deserts, and flying through the air in monoplanes. They are riding the seas in bark canoes and enormous steamships; they are painting bison and mammoths on the walls of dismal caves and covering huge canvases with queer futuristic designs. I watch the migrations from Atlantis. I watch the migrations from Lemuria. I see the elder races—a strange horde of black dwarfs overwhelming Asia and the Neandertalers with lowered heads and bent knees ranging obscenely across Europe. I watch the Achæans streaming into the Greek islands, and the crude beginnings of Hellenic culture. I am in Athens and Pericles is young. I am standing on the soil of Italy. I assist in the rape of the Sabines; I march with the Imperial Legions. I tremble with awe and wonder as the enormous standards go by and the ground shakes with the tread of the victorious *hastati*. A thousand naked slaves grovel before me as I pass in a litter of gold and ivory drawn by night-black oxen from Thebes, and the flower-girls scream *'Ave Cæsar'* as I nod and smile. I am myself a slave on a Moorish galley. I watch the erection of a great cathedral. Stone by stone it rises, and through months and years I stand and watch each stone as it falls into place. I am burned on a cross head downward in the thyme-scented gardens of Nero, and I watch with amusement and scorn the torturers at work in the chambers of the Inquisition.

"I walk in the holiest sanctuaries; I enter the temples of Venus. I kneel in adoration before the Magna Mater, and I throw coins on the bare knees of the sacred courtesans who sit with veiled faces in the groves of Babylon. I creep into an Elizabethan theatre and with the stinking rabble about me I applaud *The Merchant of Venice*. I walk with Dante through the narrow streets of Florence. I meet the young Beatrice and the hem of her garment brushes my sandals as I stare enraptured. I am a priest of Isis, and my magic astounds the nations. Simon Magus kneels before me, imploring my assistance, and Pharaoh trembles when I approach. In India I talk with the Masters and run screaming from their presence, for their revelations are as salt on wounds that bleed.

"I perceive everything *simultaneously*. I perceive everything from all sides; I am a part of all the teeming billions about me. I exist in all men and all men exist in me. I perceive the whole of human history in a single instant, the past and the present.

"By simply *straining* I can see farther and farther back. Now I am going back through strange curves and angles. Angles and curves multiply about me. I perceive great segments of time through *curves*. There is *curved time*, and *angular time*. The beings that exist in angular time can not enter curved time. It is very strange.

"I am going back and back. Man has disappeared from the earth. Gigantic reptiles crouch beneath enormous palms and swim through the loathly black waters of dismal lakes. Now the reptiles have disappeared. No animals remain upon the land, but beneath the waters, plainly visible to me, dark forms move slowly over the rotting vegetation.

"The forms are becoming simpler and simpler. Now they are single cells. All about me there are angles—strange angles that have no counterparts on the earth. I am desperately afraid.

"There is an abyss of being which man has never fathomed."

I stared. Chalmers had risen to his feet and he was gesticulating helplessly with his arms. "I am passing through *unearthly* angles; I am approaching—oh, the burning horror of it!"

"Chalmers!" I cried. "Do you wish me to interfere?"

He brought his right hand quickly before his face, as though to shut out a vision unspeakable. "Not yet!" he cried; "I will go on. I will see—what—lies—beyond—"

A cold sweat streamed from his forehead and his shoulders jerked spasmodically. "Beyond life there are"—his face grew ashen with terror—"*things* that I can not distinguish. They move slowly through angles. They have no bodies, and they move slowly through outrageous angles."

It was then that I became aware of the odour in the room. It was a pungent, indescribable odour, so nauseous that I could scarcely endure it. I stepped quickly to the window and threw it open. When I returned to Chalmers and looked into his eyes I nearly fainted.

"I think they have scented me!" he shrieked. "They are slowly turning toward me."

He was trembling horribly. For a moment he clawed at the air with his hands. Then his legs gave way beneath him and he fell forward on his face, slobbering and moaning.

I watched him in silence as he dragged himself across the floor. He was no longer a man. His teeth were bared and saliva dripped from the corners of his mouth.

"Chalmers," I cried. "Chalmers, stop it! Stop it, do you hear?"

As if in reply to my appeal he commenced to utter hoarse convulsive sounds which resembled nothing so much as the barking of a dog, and began a sort of hideous writhing in a circle about the room. I bent and seized him by the shoulders. Violently, desperately, I shook him. He turned his head and snapped at my wrist. I was sick with horror, but I dared not release him for fear that he would destroy himself in a paroxysm of rage.

"Chalmers," I muttered, "you must stop that. There is nothing in this room that can harm you. Do you understand?"

I continued to shake and admonish him, and gradually the madness died out of his face. Shivering convulsively, he crumpled into a grotesque heap on the Chinese rug.

I carried him to the sofa and deposited him upon it. His features were twisted in pain, and I knew that he was still struggling dumbly to escape from abominable memories.

"Whisky," he muttered. "You'll find a flask in the cabinet by the window—upper left-hand drawer."

When I handed him the flask his fingers tightened about it until the knuckles showed blue. "They nearly got me," he gasped. He drained the stimulant in immoderate gulps, and gradually the colour crept back into his face.

"That drug was the very devil!" I murmured.

"It wasn't the drug," he moaned.

His eyes no longer glared insanely, but he still wore the look of a lost soul.

"They scented me in time," he moaned. "I went too far."

"What were *they* like?" I said, to humour him.

He leaned forward and gripped my arm. He was shivering horribly. "No word in our language can describe them!" He spoke in a hoarse whisper. "They are symbolized vaguely in the myth of the Fall, and in an obscene form which is occasionally found engraven on ancient tablets. The Greeks had a name for them, which veiled their essential foulness. The tree, the snake and the apple—these are the vague symbols of a most awful mystery."

His voice had risen to a scream. "Frank, Frank, a terrible and unspeakable *deed* was done in the beginning. Before time, the *deed*, and from the deed—"

He had risen and was hysterically pacing the room. "The seeds of the deed move through angles in dim recesses of time. They are hungry and athirst!"

"Chalmers," I pleaded to quiet him. "We are living in the third decade of the Twentieth Century."

"They are lean and athirst!" he shrieked. *"The Hounds of Tindalos!"*

"Chalmers, shall I phone for a physician?"

"A physician can not help me now. They are horrors of the soul, and yet"—he hid his face in his hands and groaned—"they are real, Frank. I saw them for a ghastly moment. For a moment I stood on the *other side*. I stood on the pale grey shores beyond time and space. In an awful light that was not light, in a silence that shrieked, I saw *them*.

"All the evil in the universe was concentrated in their lean, hungry bodies. Or had they bodies? I saw them only for a moment; I can not be certain. *But I heard them breathe.* Indescribably for a moment I felt their breath upon my face. They turned toward me and I fled screaming. In a single moment I fled screaming through time. I fled down quintillions of years.

"But they scented me. Men awake in them cosmic hungers. We have escaped, momentarily, from the foulness that rings them round. They thirst for that in us which is clean, which emerged from the deed without stain. There is a part of us which did not partake in the deed, and that they hate. But do not imagine that they are literally, prosaically evil. They are beyond good and evil as we know it. They are that which in the beginning fell away from cleanliness. Through the deed they became bodies of death, receptacles of all foulness. But they are not evil in *our* sense because in the spheres through which they move there is no thought, no morals, no right or wrong as we understand it. There is merely the pure and the foul. The foul expresses itself through angles; the pure through curves. Man, the pure part of him, is descended from a curve. Do not laugh. I mean that literally."

I rose and searched for my hat. "I'm dreadfully sorry for you, Chalmers," I said, as I walked toward the door. "But I don't intend to stay and listen to such gibberish. I'll send my physician to see you. He's an elderly, kindly chap and he won't be offended if you tell him to go to the devil. But I hope you'll respect his advice. A week's rest in a good sanitarium should benefit you immeasurably."

I heard him laughing as I descended the stairs, but his laughter was so utterly mirthless that it moved me to tears.

II

When Chalmers phoned the following morning my first impulse was to hang up the receiver immediately. His request was so unusual and his voice was so wildly hysterical that I feared any further association with him would result in the impairment of my own sanity. But I could not doubt the genuineness of his misery, and when he broke down completely and I heard him sobbing over the wire I decided to comply with his request.

"Very well," I said. "I will come over immediately and bring the plaster."

En route to Chalmers' home I stopped at a hardware store and purchased twenty pounds of plaster of Paris. When I entered my friend's room he was crouching by the window watching the opposite wall out of eyes that were feverish with fright. When he saw me he rose and seized the parcel containing the plaster with an avidity that amazed and horrified me. He had extruded all of the furniture and the room presented a desolate appearance.

"It is just conceivable that we can thwart them!" he exclaimed. "But we must work rapidly. Frank, there is a stepladder in the hall. Bring it here immediately. And then fetch a pail of water."

"What for?" I murmured.

He turned sharply and there was a flush on his face. "To mix the plaster, you fool!" he cried. "To mix the plaster that will save our bodies and souls from a contamination unmentionable. To mix the plaster that will save the world from—Frank, *they must be kept out!*"

"Who?" I murmured.

"The Hounds of Tindalos!" he muttered. "They can only reach us through angles. We must eliminate all angles from this room. I shall plaster up all of the corners, all of the crevices. We must make this room resemble the interior of a sphere."

I knew that it would have been useless to argue with him. I fetched the stepladder, Chalmers mixed the plaster, and for three hours we laboured. We filled in the four corners of the wall and the intersections of the floor and wall and the wall and ceiling, and we rounded the sharp angles of the window-seat.

"I shall remain in this room until they return in time," he affirmed when our task was completed. "When they discover that the scent leads through curves they will return. They will return ravenous and snarling and unsatisfied to the foulness that was in the beginning, before time, beyond space."

He nodded graciously and lit a cigarette. "It was good of you to help," he said.

"Will you not see a physician, Chalmers?" I pleaded.

"Perhaps—tomorrow," he murmured. "But now I must watch and wait."

"Wait for what?" I urged.

Chalmers smiled wanly. "I know that you think me insane," he said. "You have a shrewd but prosaic mind, and you can not conceive of an entity that does not depend for its existence on force and matter. But did it ever occur to you, my friend, that force and matter are merely the barriers to perception imposed by time and space? When one knows, as I do, that time and space are identical and that they are both deceptive because they are merely imperfect manifestations of a higher reality, one no longer seeks in the visible world for an explanation of the mystery and terror of being."

I rose and walked toward the door.

"Forgive me," he cried. "I did not mean to offend you. You have a superlative intellect, but I—I have a *superhuman* one. It is only natural that I should be aware of your limitations."

"Phone if you need me," I said, and descended the stairs two steps at a time. "I'll send my physician over at once," I muttered, to myself. "He's a hopeless maniac, and heaven knows what will happen if someone doesn't take charge of him immediately."

III

The following is a condensation of two announcements which appeared in the Partridgeville Gazette *for July 3, 1928:*

EARTHQUAKE SHAKES FINANCIAL DISTRICT

At 2 o'clock this morning an earth tremor of unusual severity broke several plate-glass windows in Central Square and completely disorganized the electric and street railway systems. The tremor was felt in the outlying districts and the steeple of the First Baptist Church on Angell Hill (designed by Christopher Wren in 1717) was entirely demolished. Firemen are now attempting to put out a blaze which threatens to destroy the Partridgeville Glue Works. An investigation is promised by the mayor and an immediate attempt will be made to fix responsibility for this disastrous occurrence.

OCCULT WRITER MURDERED BY UNKNOWN GUEST

Horrible Crime in Central Square

Mystery Surrounds Death of Halpin Chalmers

At 9 a. m. today the body of Halpin Chalmers, author and journalist, was found in an empty room above the jewellery store of Smithwick and Isaacs, 24 Central Square. The coroner's investigation revealed that the room had been rented *furnished* to Mr. Chalmers on May 1, and that he had himself disposed of the furniture a fortnight ago. Chalmers was the author of several recondite books on occult themes, and a member of the Bibliographic Guild. He formerly resided in Brooklyn, New York.

At 7 a. m. Mr. L. E. Hancock, who occupies the apartment opposite Chalmers' room in the Smithwick and Isaacs establishment, smelt a peculiar odour when he opened his door to take in his cat and the morning edition of the *Partridgeville Gazette*. The odour he describes as extremely acrid and nauseous, and he affirms that it was so strong in the vicinity of Chalmers' room that he was obliged to hold his nose when he approached that section of the hall.

He was about to return to his own apartment when it occurred to him that Chalmers might have accidentally forgotten to turn off the gas in his kitchenette. Becoming considerably alarmed at the thought, he decided to investigate, and when repeated tap-pings on Chalmers' door brought no response he notified the superintendent. The latter opened the door by means of a pass key, and the two men quickly made their way into Chalmers' room. The room was utterly destitute of furniture, and Hancock asserts that when he first glanced at the floor his heart went cold within him, and that the superintendent, without saying a word, walked to the open window and stared at the building opposite for fully five minutes.

Chalmers lay stretched upon his back in the centre of the room. He was starkly nude, and his chest and arms were covered with a peculiar bluish pus or ichor. His head lay grotesquely upon his chest. It had been completely severed from his body, and the features were twisted and torn and horribly mangled. Nowhere was there a trace of blood.

The room presented a most astonishing appearance. The intersections of the walls, ceiling and floor had been thickly smeared with plaster of Paris, but at intervals fragments had cracked and fallen off, and someone had grouped these upon the floor about the murdered man so as to form a perfect triangle.

Beside the body were several sheets of charred yellow paper. These bore fantastic geometric designs and symbols and several hastily scrawled sentences. The sentences were almost illegible and so absurd in context that they furnished no possible clue to the perpetrator of the crime. "I am waiting and watching," Chalmers wrote. "I sit by the window and watch walls and ceiling. I do not believe they can reach me, but I must beware of the Doels. Perhaps *they* can help them break through. The satyrs will help, and they can advance through the scarlet circles. The Greeks knew a way of preventing that. It is a great pity that we have forgotten so much."

On another sheet of paper, the most badly charred of the seven or eight fragments found by Detective Sergeant Douglas (of the Partridgeville Reserve), was scrawled the following:

"Good God, the plaster is falling! A terrific shock has loosened the plaster and it is falling. An earthquake perhaps! I never could have anticipated this. It is growing dark in the room. I must phone Frank. But can he get here in time? I will try. I will recite the Einstein formula. I will—God, they are breaking through! They are breaking through! Smoke is pouring from the corners of the wall. Their *tongues*—ahhhhh—"

In the opinion of Detective Sergeant Douglas, Chalmers was poisoned by some obscure chemical. He has sent specimens of the strange blue slime found on Chalmers' body to the Partridgeville Chemical Laboratories; and he expects the report will shed new light on one of the most mysterious crimes of recent years. That Chalmers entertained a guest on the evening preceding the earthquake is certain, for his neighbour distinctly heard a low murmur of conversation in the former's room as he passed it on his way to the stairs. Suspicion points strongly to this unknown visitor and the police are diligently endeavouring to discover his identity.

IV

Report of James Morton, chemist and bacteriologist:

My dear Mr. Douglas:

The fluid sent to me for analysis is the most peculiar that I have ever examined. It resembles living protoplasm, but it lacks the peculiar substances known as enzymes. Enzymes catalyse the chemical reactions occurring in living cells, and when the cell dies they cause it to disintegrate by hydrolysation. Without enzymes protoplasm should possess enduring vitality, i. e., immortality. Enzymes are the negative components, so to speak, of unicellular organism, which is the basis of all life. That living matter can exist without enzymes biologists emphatically deny. And yet the substance that you have sent me is alive and it lacks these "indispensable" bodies. Good God, sir, do you realize what astounding new vistas this opens up?

V

Excerpt from The Secret Watchers *by the late Halpin Chalmers:*

What if, parallel to the life we know, there is another life that does not die, which lacks the elements that destroy *our* life? Perhaps in another dimension there is a *different* force from that which generates our life. Perhaps this force emits energy, or something similar to energy, which passes from the unknown dimension where *it* is and creates a new form of cell life in our dimension. No one knows that such new cell life does exist in our dimension. Ah, but I have seen *its* manifestations. I have *talked* with them. In my room at night I have talked with the Doels. And in dreams I have seen their maker. I have stood on the dim shore beyond time and matter and seen *it*. *It* moves through strange curves and outrageous angles. Some day I shall travel in time and meet *it* face to face.

—1932—

THE TRAP

Henry S. Whitehead and H. P. Lovecraft

Relatively unknown in his own lifetime, H. P. Lovecraft (1890–1937) is now the most famous and influential of all weird fiction authors, and perhaps the principal reason for the genre's survival in the second half of the twentieth century. Weird fiction's ongoing Renaissance, however, has put Lovecraft centre-stage once more, and for different reasons. As revelations about his racism and xenophobia reach a wider audience, Lovecraft's status, and the shadow he casts over weird fiction, has been the subject of much debate. This particular story was co-written with Henry S. Whitehead (1882–1932). A deacon in the Episcopal Church, Whitehead produced almost fifty stories for pulps like Weird Tales *and* Strange Tales. *This story was first published in the latter in March 1932, and warns of the dangers of old mirrors.*

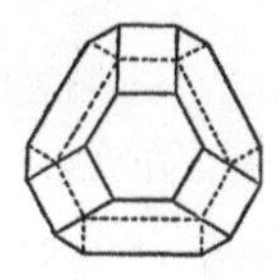

IT WAS ON A CERTAIN THURSDAY MORNING IN DECEMBER THAT the whole thing began with that unaccountable motion I thought I saw in my antique Copenhagen mirror. Something, it seemed to me, stirred—something reflected in the glass, though I was alone in my quarters. I paused and looked intently, then, deciding that the effect must be a pure illusion, resumed the interrupted brushing of my hair.

I had discovered the old mirror, covered with dust and cobwebs, in an outbuilding of an abandoned estate-house in Santa Cruz's sparsely settled Northside territory, and had brought it to the United States from the Virgin Islands. The venerable glass was dim from more than two hundred years' exposure to a tropical climate, and the graceful ornamentation along the top of the gilt frame had been badly smashed. I had had the detached pieces set back into the frame before placing it in storage with my other belongings.

Now, several years later, I was staying half as a guest and half as a tutor at the private school of my old friend Browne on a windy Connecticut hillside—occupying an unused wing in one of the dormitories, where I had two rooms and a hallway to myself. The old mirror, stowed securely in mattresses, was the first of my possessions to be unpacked on my arrival; and I had set it up majestically in the living-room, on top of an old rosewood console which had belonged to my great-grandmother.

The door of my bedroom was just opposite that of the living-room, with a hallway between; and I had noticed that by looking into my chiffonier glass I could see the larger mirror through the two doorways—which was exactly like glancing down an endless, though diminishing, corridor. On this Thursday morning I thought I saw a curious suggestion of motion down that normally empty corridor—but, as I have said, soon dismissed the notion.

When I reached the dining-room I found everyone complaining of the cold, and learned that the school's heating-plant was temporarily out of order. Being especially sensitive to low temperatures, I was myself an acute sufferer; and at once decided not to brave any freezing schoolroom that day. Accordingly I invited my class to come over to my living-room for an informal session around my grate-fire—a suggestion which the boys received enthusiastically.

After the session one of the boys, Robert Grandison, asked if he might remain; since he had no appointment for the second morning period. I told him to stay, and welcome. He sat down to study in front of the fireplace in a comfortable chair.

It was not long, however, before Robert moved to another chair somewhat farther away from the freshly replenished blaze, this change bringing him directly opposite the old mirror. From my own chair in another part of the room I noticed how fixedly he began to look at the dim, cloudy glass, and, wondering what so greatly interested him, was reminded of my own experience earlier that morning. As time passed he continued to gaze, a slight frown knitting his brows.

At last I quietly asked him what had attracted his attention. Slowly, and still wearing the puzzled frown, he looked over and replied rather cautiously:

"It's the corrugations in the glass—or whatever they are, Mr. Canevin. I was noticing how they all seem to run from a certain point. Look—I'll show you what I mean."

The boy jumped up, went over to the mirror, and placed his finger on a point near its lower left-hand corner.

"It's right here, sir," he explained, turning to look toward me and keeping his finger on the chosen spot.

His muscular action in turning may have pressed his finger against the glass. Suddenly he withdrew his hand as though with some slight effort, and with a faintly muttered "Ouch." Then he looked back at the glass in obvious mystification.

"What happened?" I asked, rising and approaching.

"Why—it—" He seemed embarrassed. "It—I—felt—well, as though it were pulling my finger into it. Seems—er—perfectly foolish, sir, but—well—it was a most peculiar sensation." Robert had an unusual vocabulary for his fifteen years.

I came over and had him show me the exact spot he meant.

"You'll think I'm rather a fool, sir," he said shamefacedly, "but—well, from right here I can't be absolutely sure. From the chair it seemed to be clear enough."

Now thoroughly interested, I sat down in the chair Robert had occupied and looked at the spot he selected on the mirror. Instantly the thing "jumped out at me." Unmistakably, from that particular angle, all the many whorls in the ancient glass appeared to converge like a large number of spread strings held in one hand and radiating out in streams.

Getting up and crossing to the mirror, I could no longer see the curious spot. Only from certain angles, apparently, was it visible.

Directly viewed, that portion of the mirror did not even give back a normal reflection—for I could not see my face in it. Manifestly I had a minor puzzle on my hands.

Presently the school gong sounded, and the fascinated Robert Grandison departed hurriedly, leaving me alone with my odd little problem in optics. I raised several window-shades, crossed the hallway, and sought for the spot in the chiffonier mirror's reflection. Finding it readily, I looked very intently and thought I again detected something of the "motion." I craned my neck, and at last, at a certain angle of vision, the thing again "jumped out at me."

The vague "motion" was now positive and definite—an appearance of torsional movement, or of whirling; much like a minute yet intense whirlwind or waterspout, or a huddle of autumn leaves dancing circularly in an eddy of wind along a level lawn. It was, like the earth's, a double motion—around and around, and at the same time *inward*, as if the whorls poured themselves endlessly toward some point inside the glass. Fascinated, yet realizing that the thing must be an illusion, I grasped an impression of quite distinct *suction*, and thought of Robert's embarrassed explanation: "*I felt as though it were pulling my finger into it.*"

A kind of slight chill ran suddenly up and down my backbone. There was something here distinctly worth looking into. And as the idea of investigation came to me, I recalled the rather wistful expression of Robert Grandison when the gong called him to class. I remembered how he had looked back over his shoulder as he walked obediently out into the hallway, and resolved that he should be included in whatever analysis I might make of this little mystery.

⋆

Exciting events connected with that same Robert, however, were soon to chase all thoughts of the mirror from my consciousness for a time. I was away all that afternoon, and did not return to the school until the five-fifteen "Call-over"—a general assembly at which the boys' attendance was compulsory. Dropping in at this function with the idea of picking Robert up for a session with the mirror, I was astonished and pained to find him absent—a very unusual and unaccountable thing in his case. That evening Browne told me that the boy had actually disappeared, a search in his room, in the gymnasium, and in all other accustomed places being unavailing, though all his belongings—including his outdoor clothing—were in their proper places.

He had not been encountered on the ice or with any of the hiking groups that afternoon, and telephone calls to all the school-catering merchants of the neighbourhood were in vain. There was, in short, no record of his having been seen since the end of the lesson periods at two-fifteen; when he had turned up the stairs toward his room in Dormitory Number Three.

When the disappearance was fully realized, the resulting sensation was tremendous throughout the school. Browne, as headmaster, had to bear the brunt of it; and such an unprecedented occurrence in his well-regulated, highly-organized institution left him quite bewildered. It was learned that Robert had not run away to his home in western Pennsylvania, nor did any of the searching-parties of boys and masters find any trace of him in the snowy countryside around the school. So far as could be seen, he had simply vanished.

Robert's parents arrived on the afternoon of the second day after his disappearance. They took their trouble quietly, though, of course, they were staggered by this unexpected disaster. Browne looked ten years older for it, but there was absolutely nothing that could be done. By the fourth day the case had settled down in the opinion of the school as an insoluble mystery. Mr. and Mrs. Grandison went reluctantly back to their home, and on the following morning the ten days' Christmas vacation began.

Boys and masters departed in anything but the usual holiday spirit; and Browne and his wife were left, along with the servants, as my only fellow-occupants of the big place. Without the masters and boys it seemed a very hollow shell indeed.

* * *

That afternoon I sat in front of my grate-fire thinking about Robert's disappearance and evolving all sorts of fantastic theories to account for it. By evening I had acquired a bad headache, and ate a light supper accordingly. Then, after a brisk walk around the massed buildings, I returned to my living-room and took up the burden of thought once more.

A little after ten o'clock I awakened in my armchair, stiff and chilled, from a doze during which I had let the fire go out. I was physically uncomfortable, yet mentally aroused by a peculiar sensation of expectancy and possible hope. Of course it had to do with the problem that was harassing me. For I had started from that inadvertent nap with a curious, persistent idea—the odd idea that a tenuous, hardly recognizable Robert Grandison had been trying desperately to communicate with me. I finally went to bed with one conviction unreasoningly strong in my mind. Somehow I was sure that young Robert Grandison was still alive.

That I should be receptive of such a notion will not seem strange to those who know of my long residence in the West Indies and my close contact with unexplained happenings there. It will not seem strange, either, that I fell asleep with an urgent desire to establish some sort of mental communication with the missing boy. Even the most prosaic scientists affirm, with Freud, Jung, and Adler, that the subconscious mind is most open to external impression in sleep; though such impressions are seldom carried over intact into the waking state.

Going a step further and granting the existence of telepathic forces, it follows that such forces must act most strongly on a sleeper; so that if I were ever to get a definite message from Robert, it would be during a period of profoundest slumber. Of course, I might lose the message in waking; but my aptitude for retaining such things has been sharpened by types of mental discipline picked up in various obscure corners of the globe.

I must have dropped asleep instantaneously, and from the vividness of my dreams and the absence of wakeful intervals I judge that my sleep was a very deep one. It was six forty-five when I awakened, and there still lingered with me certain impressions which I knew were carried over from the world of somnolent cerebration. Filling my mind was the vision of Robert Grandison strangely transformed to a boy of a dull greenish dark-blue colour; Robert desperately endeavouring to communicate with me by means of speech, yet finding some almost insuperable difficulty in so doing. A wall of curious spatial separation seemed to stand between him and me—a mysterious, invisible wall which completely baffled us both.

I had seen Robert as though at some distance, yet queerly enough he seemed at the same time to be just beside me. He was both larger and smaller than in real life, his apparent size varying *directly*, instead of *inversely*, with the distance as he advanced and retreated in the course of conversation. That is, he grew larger instead of smaller to my eye when he stepped away or backwards, and vice versa; as if the laws of perspective in his case had been wholly reversed. His aspect was misty and uncertain—as if he lacked sharp or permanent outlines; and the anomalies of his colouring and clothing baffled me utterly at first.

At some point in my dream Robert's vocal efforts had finally crystallized into audible speech—albeit speech of an abnormal thickness and dullness. I could not for a time understand anything he said, and even in the dream racked my brain for a clue to where he was, what he wanted to tell, and why his utterance was so clumsy and unintelligible. Then little by little I began to distinguish words and phrases, the very first of which sufficed to throw my dreaming self into the wildest excitement and to establish a certain mental connection which had previously refused to take conscious form because of the utter incredibility of what it implied.

I do not know how long I listened to those halting words amidst my deep slumber, but hours must have passed while the strangely remote speaker struggled on with his tale. There was revealed to me such a circumstance as I cannot hope to make others believe without the strongest corroborative evidence, yet which I was quite ready to accept as truth—both in the dream and after waking—because of my formed contacts with uncanny things. The boy was obviously watching my face—mobile in receptive sleep—as he choked along; for about the time I began to comprehend him, his own expression brightened and gave signs of gratitude and hope.

Any attempt to hint at Robert's message, as it lingered in my ears after a sudden awakening in the cold, brings this narrative to a point where I must choose my words with the greatest care. Everything involved is so difficult to record that one tends to flounder helplessly. I have said that the revelation established in my mind a certain connection which reason had not allowed me to formulate consciously before. This connection, I need no longer hesitate to hint, had to do with the old Copenhagen mirror whose suggestions of motion had so impressed me on the morning of the disappearance, and whose whorl-like contours and apparent illusions of suction had later exerted such a disquieting fascination on both Robert and me.

Resolutely, though my outer consciousness had previously rejected what my intuition would have liked to imply, it could reject that stupendous conception no longer. What was fantasy in the tale of "Alice" now came to me as a grave and immediate reality. That looking-glass had indeed possessed a malign, abnormal suction; and the struggling speaker in my dream made clear the extent to which it violated all the known precedents of human experience and all the age-old laws of our three sane dimensions. It was more than a mirror—it was a gate; a trap; a link with spatial recesses not meant for the denizens of our visible universe, and realizable only in terms of the most intricate non-Euclidean mathematics. *And in some outrageous fashion Robert Grandison had passed out of our ken into the glass and was there immured, waiting for release.*

It is significant that upon awakening I harboured no genuine doubt of the reality of the revelation. That I had actually held conversation with a trans-dimensional Robert, rather than evoked the whole episode from my broodings about his disappearance and about the old illusions of the mirror, was as certain to my inmost instincts as any of the instinctive certainties commonly recognized as valid.

The tale thus unfolded to me was of the most incredibly bizarre character. As had been clear on the morning of his disappearance, Robert was intensely fascinated by the ancient mirror. All through the hours of school, he had it in mind to come back to my living-room and examine it further. When he did arrive, after the close of the school day, it was somewhat later than two-twenty, and I was absent in town. Finding me out and knowing that I would not mind, he had come into my living-room and gone straight to the mirror; standing before it and studying the place where, as we had noted, the whorls appeared to converge.

Then, quite suddenly, there had come to him an overpowering urge to place his hand upon this whorl-centre. Almost reluctantly, against his better judgment, he had done so; and upon making the contact had felt at once the strange, almost painful suction which had perplexed him that morning. Immediately thereafter—quite without warning, but with a wrench which seemed to twist and tear every bone and muscle in his body and to bulge and press and cut at every nerve—he had been abruptly *drawn through* and found himself *inside.*

Once through, the excruciatingly painful stress upon his entire system was suddenly released. He felt, he said, as though he had just been born—a feeling that made itself evident every time he tried to do anything; walk, stoop, turn his head, or utter speech. Everything about his body seemed a misfit.

These sensations wore off after a long while, Robert's body becoming an organized whole rather than a number of protesting parts. Of all the forms of expression, speech remained the most difficult; doubtless because it is complicated, bringing into play a number of different organs, muscles, and tendons. Robert's feet, on the other hand, were the first members to adjust themselves to the new conditions within the glass.

During the morning hours I rehearsed the whole reason-defying problem; correlating everything I had seen and heard, dismissing the natural scepticism of a man of sense, and scheming to devise possible plans for Robert's release from his incredible prison. As I did so a number of originally perplexing points became clear—or at least, clearer—to me.

There was, for example, the matter of Robert's colouring. His face and hands, as I have indicated, were a kind of dull greenish dark-blue; and I may add that his familiar blue Norfolk jacket had turned to a pale lemon-yellow while his trousers remained a neutral grey as before. Reflecting on this after waking, I found the circumstance closely allied to the reversal of perspective which made Robert seem to grow larger when receding and smaller when approaching. Here, too, was a physical *reversal*—for every detail of his colouring in the unknown dimension was the exact reverse or complement of the corresponding colour detail in normal life. In physics the typical complementary colours are blue and yellow, and red and green. These pairs are opposites, and when mixed yield grey. Robert's natural colour was a pinkish-buff, the opposite of which is the greenish-blue I saw. His blue coat had become yellow, while the grey trousers remained grey. This latter point baffled me until I remembered that grey is itself a mixture of opposites. There is no opposite for grey—or rather, it is its own opposite.

Another clarified point was that pertaining to Robert's curiously dulled and thickened speech—as well as to the general awkwardness and sense of misfit bodily parts of which he had complained. This, at the outset, was a puzzle indeed; though after long thought the clue occurred to me. Here again was the same *reversal* which affected perspective and colouration. Anyone in the fourth dimension must necessarily be reversed in just this way—hands and feet, as well as colours and perspectives, being changed about. It would be the same with all the other dual organs, such as nostrils, ears, and eyes. Thus Robert had been talking with a reversed tongue, teeth, vocal cords, and kindred speech-apparatus; so that his difficulties in utterance were little to be wondered at.

As the morning wore on, my sense of the stark reality and maddening urgency of the dream-disclosed situation increased rather than decreased. More and more I felt that something must be done, yet realized that I could not seek advice or aid. Such a story as mine—a conviction based upon mere dreaming—could not conceivably bring me anything but ridicule or suspicions as to my mental state. And what, indeed, could I do, aided or unaided, with as little working data as my nocturnal impressions had provided? I must, I finally recognized, have more information before I could even think of a possible plan for releasing Robert. This could come only through the receptive conditions of sleep, and it heartened me to reflect that according to every probability my telepathic contact would be resumed the moment I fell into deep slumber again.

I accomplished sleeping that afternoon, after a midday dinner at which, through rigid self-control, I succeeded in concealing from Browne and his wife the tumultuous thoughts that crashed through my mind. Hardly had my eyes closed when a dim telepathic image began to appear; and I soon realized to my infinite excitement that it was identical with what I had seen before. If anything, it was more distinct; and when it began to speak I seemed able to grasp a greater proportion of the words.

During this sleep I found most of the morning's deductions confirmed, though the interview was mysteriously cut off long prior to my awakening. Robert had seemed apprehensive just before communication ceased, but had already told me that in his strange fourth-dimensional prison colours and spatial relationships were indeed reversed—black being white, distance increasing apparent size, and so on.

He had also intimated that, notwithstanding his possession of full physical form and sensations, most human vital properties seemed curiously suspended. Nutriment, for example, was quite unnecessary—a phenomenon really more singular than the omnipresent reversal of objects and attributes, since the latter was a reasonable and mathematically indicated state of things. Another significant piece of information was that the only exit from the glass to the world was the entrance-way, and that this was permanently barred and impenetrably sealed, so far as egress was concerned.

That night I had another visitation from Robert; nor did such impressions, received at odd intervals while I slept receptively-minded, cease during the entire period of his incarceration. His efforts to communicate were desperate and often pitiful; for at times the telepathic bond would weaken, while at other times fatigue, excitement, or fear of interruption would hamper and thicken his speech.

*

I may as well narrate as a continuous whole all that Robert told me throughout the whole series of transient mental contacts—perhaps supplementing it at certain points with facts directly related after his release. The telepathic information was fragmentary and often nearly inarticulate, but I studied it over and over during the waking intervals of three intense days; classifying and cogitating with feverish diligence, since it was all that I had to go upon if the boy were to be brought back into our world.

The fourth-dimensional region in which Robert found himself was not, as in scientific romance, an unknown and infinite realm of strange sights and fantastic denizens; but was rather a projection of certain limited parts of our own terrestrial sphere within an alien and normally inaccessible aspect or direction of space. It was a curiously fragmentary, intangible, and heterogeneous world—a series of apparently dissociated scenes merging indistinctly one into the other; their constituent details having an obviously different status from that of an object drawn into the ancient mirror as Robert had been drawn. These scenes were like dream-vistas or magic-lantern images—elusive visual impressions of which the boy was not really a part, but which formed a sort of panoramic background or ethereal environment against which or amidst which he moved.

He could not touch any of the parts of these scenes—walls, trees, furniture, and the like—but whether this was because they were truly non-material, or because they always receded at his approach, he was singularly unable to determine. Everything seemed fluid, mutable, and unreal. When he walked, it appeared to be on whatever lower surface the visible scene might have—floor, path, greensward, or such; but upon analysis he always found that the contact was an illusion. There was never any difference in the resisting force met by his feet—and by his hands when he would stoop experimentally—no matter what changes of apparent surface might be involved. He could not describe this foundation or limiting plane on which he walked as anything more definite than a virtually abstract pressure balancing his gravity. Of definite tactile distinctiveness it had none, and supplementing it there seemed to be a kind of restricted levitational force which accomplished transfers of altitude. He could never actually climb stairs, yet would gradually walk up from a lower level to a higher.

Passage from one definite scene to another involved a sort of gliding through a region of shadow or blurred focus where the details of each scene mingled curiously. All the vistas were distinguished by the absence of transient objects, and the indefinite or ambiguous appearance of such semi-transient objects as furniture or details of vegetation. The lighting of every scene was diffuse and perplexing, and of course the scheme of reversed colours—bright red grass, yellow sky with confused black and grey cloud-forms, white tree-trunks, and green brick walls—gave to everything an air of unbelievable grotesquerie. There was an alteration of day and night, which turned out to be a reversal of the normal hours of light and darkness at whatever point on the earth the mirror might be hanging.

This seemingly irrelevant diversity of the scenes puzzled Robert until he realized that they comprised merely such places as had been reflected for long continuous periods in the ancient glass. This also explained the odd absence of transient objects, the generally arbitrary boundaries of vision, and the fact that all exteriors were framed by the outlines of doorways or windows. The glass, it appeared, had power to store up these intangible scenes through long exposure; though it could never absorb anything corporeally, as Robert had been absorbed, except by a very different and particular process.

But—to me at least—the most incredible aspect of the mad phenomenon was the monstrous subversion of our known laws of space involved in the relation of the various illusory scenes to the actual terrestrial regions represented. I have spoken of the glass as storing up the images of these regions, but this is really an inexact definition. In truth, each of the mirror scenes formed a true and quasi-permanent fourth-dimensional projection of the corresponding mundane region; so that whenever Robert moved to a certain part of a certain scene, as he moved into the image of my room when sending his telepathic messages, *he was actually in that place itself, on earth*—though under spatial conditions which cut off all sensory communication, in either direction, between him and the present tri-dimensional aspect of the place.

Theoretically speaking, a prisoner in the glass could in a few moments go anywhere on our planet—into any place, that is, which had ever been reflected in the mirror's surface. This probably applied even to places where the mirror had not hung long enough to produce a clear illusory scene; the terrestrial region being then represented by a zone of more or less formless shadow. Outside the definite scenes was a seemingly limitless waste of neutral grey shadow about which Robert could never be certain, and into which he never dared stray far lest he become hopelessly lost to the real and mirror worlds alike.

Among the earliest particulars which Robert gave, was the fact that he was not alone in his confinement. Various others, all in antique garb, were in there with him—a corpulent middle-aged gentleman with tied queue and velvet knee-breeches who spoke English fluently though with a marked Scandinavian accent; a rather beautiful small girl with very blonde hair which appeared as glossy dark blue; two apparently mute black men whose features contrasted with the pallor of their reversed-coloured skins; three young men; one young woman; a very small child, almost an infant; and a lean, elderly Dane of extremely distinctive aspect and a kind of half-malign intellectuality of countenance.

This last named individual—Axel Holm, who wore the satin small-clothes, flared-skirted coat, and voluminous full-bottomed periwig of an age more than two centuries in the past—was notable among the little band as being the one responsible for the presence of them all. He it was who, skilled equally in the arts of magic and glass working, had long ago fashioned this strange dimensional prison in which himself, his slaves, and those whom he chose to invite or allure thither were immured unchangingly for as long as the mirror might endure.

Holm was born early in the seventeenth century, and had followed with tremendous competence and success the trade of a glass-blower and moulder in Copenhagen. His glass, especially in the form of large drawing-room mirrors, was always at a premium. But the same bold mind which had made him the first glazier of Europe also served to carry his interests and ambitions far beyond the sphere of mere material craftsmanship. He had studied the world around him, and chafed at the limitations of human knowledge and capability. Eventually he sought for dark ways to overcome those limitations, and gained more success than is good for any mortal.

He had aspired to enjoy something like eternity, the mirror being his provision to secure this end. Serious study of the fourth dimension was far from beginning with Einstein in our own era; and Holm, more than erudite in all the methods of his day, knew that a bodily entrance into that hidden phase of space would prevent him from dying in the ordinary physical sense. Research showed him that the principle of reflection undoubtedly forms the chief gate to all dimensions beyond our familiar three; and chance placed in his hands a small and very ancient glass whose cryptic properties he believed he could turn to advantage. Once "inside" this mirror according to the method he had envisaged, he felt that "life" in the sense of form and consciousness would go on virtually forever, provided the mirror could be preserved indefinitely from breakage or deterioration.

Holm made a magnificent mirror, such as would be prized and carefully preserved; and in it deftly fused the strange whorl-configured relic he had acquired. Having thus prepared his refuge and his trap, he began to plan his mode of entrance and conditions of tenancy. He would have with him both servitors and companions; and as an experimental beginning he sent before him into the glass two dependable slaves brought from the West Indies. What his sensations must have been upon beholding this first concrete demonstration of his theories, only imagination can conceive.

Undoubtedly a man of his knowledge realized that absence from the outside world if deferred beyond the natural span of life of those within, must mean instant dissolution at the first attempt to return to that world. But, barring that misfortune or accidental breakage, those within would remain forever as they were at the time of entrance. They would never grow old, and would need neither food nor drink.

To make his prison tolerable he sent ahead of him certain books and writing materials, a chair and table of stoutest workmanship, and a few other accessories. He knew that the images which the glass would reflect or absorb would not be tangible, but would merely extend around him like a background of dream. His own transition in 1687 was a momentous experience; and must have been attended by mixed sensations of triumph and terror. Had anything gone wrong, there were frightful possibilities of being lost in dark and inconceivable multiple dimensions.

For over fifty years he had been unable to secure any additions to the little company of himself and slaves, but later on he had perfected his telepathic method of visualizing small sections of the outside world close to the glass, and attracting certain individuals in those areas through the mirror's strange entrance. Thus Robert, influenced into a desire to press upon the "door," had been lured within. Such visualizations depended wholly on telepathy, since no one inside the mirror could see out into the world of men.

It was, in truth, a strange life that Holm and his company had lived inside the glass. Since the mirror had stood for fully a century with its face to the dusty stone wall of the shed where I found it, Robert was the first being to enter this limbo after all that interval. His arrival was a gala event, for he brought news of the outside world which must have been of the most startling impressiveness to the more thoughtful of those within. He, in his turn—young though he was—felt overwhelmingly the weirdness of meeting and talking with persons who had been alive in the seventeenth and eighteenth centuries.

*

The deadly monotony of life for the prisoners can only be vaguely conjectured. As mentioned, its extensive spatial variety was limited to localities which had been reflected in the mirror for long periods; and many of these had become dim and strange as tropical climates had made inroads on the surface. Certain localities were bright and beautiful, and in these the company usually gathered. But no scene could be fully satisfying; since the visible objects were all unreal and intangible, and often of perplexingly indefinite outline. When the tedious periods of darkness came, the general custom was to indulge in memories, reflections, or conversations. Each one of that strange, pathetic group had retained his or her personality unchanged and unchangeable, since becoming immune to the time effects of outside space.

The number of inanimate objects within the glass, aside from the clothing of the prisoners, was very small; being largely limited to the accessories Holm had provided for himself. The rest did without even furniture, since sleep and fatigue had vanished along with most other vital attributes. Such inorganic things as were present, seemed as exempt from decay as the living beings. The lower forms of animal life were wholly absent.

Robert derived most of his information from Herr Thiele, the gentleman who spoke English with a Scandinavian accent. This portly Dane had taken a fancy to him, and talked at considerable length. The others, too, had received him with courtesy and goodwill; Holm himself, seeming well-disposed, had told him about various matters including the door of the trap.

The boy, as he told me later, was sensible enough never to attempt communication with me when Holm was nearby. Twice, while thus engaged, he had seen Holm appear; and had accordingly ceased at once. At no time could I see the world behind the mirror's surface. Robert's visual image, which included his bodily form and the clothing connected with it, was—like the aural image of his halting voice and like his own visualization of myself—a case of purely telepathic transmission; and did not involve true inter-dimensional sight. However, had Robert been as trained a telepathist as Holm, he might have transmitted a few strong images apart from his immediate person.

Throughout this period of revelation I had, of course, been desperately trying to devise a method for Robert's release. On the fourth day—the ninth after the disappearance—I hit on a solution. Everything considered, my laboriously formulated process was not a very complicated one; though I could not tell beforehand how it would work, while the possibility of ruinous consequences in case of a slip was appalling. This process depended, basically, on the fact that there was no possible exit from inside the glass. If Holm and his prisoners were permanently sealed in, then release must come wholly from outside. Other considerations included the disposal of the other prisoners, if any survived, and especially of Axel Holm. What Robert had told me of him was anything but reassuring; and I certainly did not wish him loose in my apartment, free once more to work his evil will upon the world. The telepathic messages had not made fully clear the effect of liberation on those who had entered the glass so long ago.

There was, too, a final though minor problem in case of success—that of getting Robert back into the routine of school life without having to explain the incredible. In case of failure, it was highly inadvisable to have witnesses present at the release operations—and lacking these, I simply could not attempt to relate the actual facts if I should succeed. Even to me the reality seemed a mad one whenever I let my mind turn from the data so compellingly presented in that tense series of dreams.

When I had thought these problems through as far as possible, I procured a large magnifying-glass from the school laboratory and studied minutely every square millimetre of that whorl-centre which presumably marked the extent of the original ancient mirror used by Holm. Even with this aid I could not quite trace the exact boundary between the old area and the surface added by the Danish wizard; but after a long study decided on a conjectural oval boundary which I outlined very precisely with a soft blue pencil, I then made a trip to Stamford, where I procured a heavy glass-cutting tool; for my primary idea was to remove the ancient and magically potent mirror from its later setting.

My next step was to figure out the best time of day to make the crucial experiment. I finally settled on two-thirty A.M.—both because it was a good season for uninterrupted work, and because it was the "opposite" of two-thirty P.M., the probable moment at which Robert had entered the mirror. This form of "oppositeness" may or may not have been relevant, but I knew at least that the chosen hour was as good as any—and perhaps better than most.

I finally set to work in the early morning of the eleventh day after the disappearance, having drawn all the shades of my living-room and closed and locked the door into the hallway. Following with breathless care the elliptical line I had traced, I worked around the whorl-section with my steel-wheeled cutting tool. The ancient glass, half an inch thick, crackled crisply under the firm, uniform pressure; and upon completing the circuit I cut around it a second time, crunching the roller more deeply into the glass.

Then, very carefully indeed, I lifted the heavy mirror down from its console and leaned it face-inward against the wall; prying off two of the thin, narrow boards nailed to the back. With equal caution I smartly tapped the cut-around space with the heavy wooden handle of the glass-cutter.

At the very first tap the whorl-containing section of glass dropped out on the Bokhara rug beneath. I did not know what might happen, but was keyed up for anything, and took a deep involuntary breath. I was on my knees for convenience at the moment, with my face quite near the newly made aperture; and as I breathed there poured into my nostrils a powerful *dusty* odour—a smell not comparable to any other I have ever encountered. Then everything within my range of vision suddenly turned to a dull grey before my failing eyesight as I felt myself overpowered by an invisible force which robbed my muscles of their power to function.

I remember grasping weakly and futilely at the edge of the nearest window drapery and feeling it rip loose from its fastening. Then I sank slowly to the floor as the darkness of oblivion passed over me.

When I regained consciousness I was lying on the Bokhara rug with my legs held unaccountably up in the air. The room was full of that hideous and inexplicable dusty smell—and as my eyes began to take in definite images I saw that Robert Grandison stood in front of me. It was he—fully in the flesh and with his colouring normal—who was holding my legs aloft to bring the blood back to my head as the school's first-aid course had taught him to do with persons who had fainted. For a moment I was struck mute by the stifling odour and by a bewilderment which quickly merged into a sense of triumph. Then I found myself able to move and speak collectedly.

I raised a tentative hand and waved feebly at Robert.

"All right, old man," I murmured, "you can let my legs down now. Many thanks. I'm all right again, I think. It was the smell—I imagine—that got me. Open that farthest window, please—wide—from the bottom. That's it—thanks. No—leave the shade down the way it was."

I struggled to my feet, my disturbed circulation adjusting itself in waves, and stood upright hanging to the back of a big chair. I was still "groggy," but a blast of fresh, bitterly cold air from the window revived me rapidly. I sat down in the big chair and looked at Robert, now walking toward me.

"First," I said hurriedly, "tell me, Robert—those others—Holm? What happened to *them*, when I—opened the exit?"

Robert paused half-way across the room and looked at me very gravely.

"I saw them fade away—into nothingness—Mr. Canevin," he said with solemnity; "and with them—everything. There isn't any more 'inside,' sir—thank God, and you, sir!"

And young Robert, at last yielding to the sustained strain which he had borne through all those terrible eleven days, suddenly broke down like a little child and began to weep hysterically in great, stifling, dry sobs.

I picked him up and placed him gently on my davenport, threw a rug over him, sat down by his side, and put a calming hand on his forehead.

"Take it easy, old fellow," I said soothingly.

The boy's sudden and very natural hysteria passed as quickly as it had come on as I talked to him reassuringly about my plans for his quiet restoration to the school. The interest of the situation and the need of concealing the incredible truth beneath a rational explanation took hold of his imagination as I had expected; and at last he sat up eagerly, telling the details of his release and listening to the instructions I had thought out. He had, it seems, been in the "projected area" of my bedroom when I opened the way back, and had emerged in that actual room—hardly realizing that he was "out." Upon hearing a fall in the living-room he had hastened thither, finding me on the rug in my fainting spell.

I need mention only briefly my method of restoring Robert in a seemingly normal way—how I smuggled him out of the window in an old hat and sweater of mine, took him down the road in my quietly started car, coached him carefully in a tale I had devised, and returned to arouse Browne with the news of his discovery. He had, I explained, been walking alone on the afternoon of his disappearance; and had been offered a motor ride by two young men who, as a joke and over his protests that he could go no farther than Stamford and back, had begun to carry him past that town. Jumping from the car during a traffic stop with the intention of hitch-hiking back before Call-Over, he had been hit by another car just as the traffic was released—awakening ten days later in the Greenwich home of the people who had hit him. On learning the date, I added, he had immediately telephoned the school; and I, being the only one awake, had answered the call and hurried after him in my car without stopping to notify anyone.

*

Browne, who at once telephoned to Robert's parents, accepted my story without question; and forbore to interrogate the boy because of the latter's manifest exhaustion. It was arranged that he should remain at the school for a rest, under the expert care of Mrs. Browne, a former trained nurse. I naturally saw a good deal of him during the remainder of the Christmas vacation, and was thus enabled to fill in certain gaps in his fragmentary dream-story.

Now and then we would almost doubt the actuality of what had occurred; wondering whether we had not both shared some monstrous delusion born of the mirror's glittering hypnotism, and whether the tale of the ride and accident were not after all the real truth. But whenever we did so we would be brought back to belief by some monstrous and haunting memory; with me, of Robert's dream-figure and its thick voice and inverted colours; with him, of the whole fantastic pageantry of ancient people and dead scenes that he had witnessed. And then there was the joint recollection of that damnable dusty odour... We knew what it meant: the instant dissolution of those who had entered an alien dimension a century and more ago.

There are, in addition, at least two lines of rather more positive evidence; one of which comes through my researches in Danish annals concerning the sorcerer, Axel Holm. Such a person, indeed, left many traces in folklore and written records; and diligent library sessions, plus conferences with various learned Danes, have shed much light on his evil fame. At present I need say only that the Copenhagen glass-blower—born in 1612—was a notorious Luciferian whose pursuits and final vanishing formed a matter of awed debate over two centuries ago. He had burned with a desire to know all things and to conquer every limitation of mankind—to which end he had delved deeply into occult and forbidden fields ever since he was a child.

He was commonly held to have joined a coven of the dreaded witchcult, and the vast lore of ancient Scandinavian myth—with its Loki the Sly One and the accursed Fenris-Wolf—was soon an open book to him. He had strange interests and objectives, few of which were definitely known, but some of which were recognized as intolerably evil. It is recorded that his two helpers, originally slaves from the Danish West Indies, had become mute soon after their acquisition by him; and that they had disappeared not long before his own disappearance from the ken of mankind.

Near the close of an already long life the idea of a glass of immortality appears to have entered his mind. That he had acquired an enchanted mirror of inconceivable antiquity was a matter of common whispering; it being alleged that he had purloined it from a fellow-sorcerer who had entrusted it to him for polishing.

This mirror—according to popular tales a trophy as potent in its way as the better-known Aegis of Minerva or Hammer of Thor—was a small oval object called "Loki's Glass," made of some polished fusible mineral and having magical properties which included the divination of the immediate future and the power to show the possessor his enemies. That it had deeper potential properties, realizable in the hands of an erudite magician, none of the common people doubted; and even educated persons attached much fearful importance to Holm's rumoured attempts to incorporate it in a larger glass of immortality. Then had come the wizard's disappearance in 1687, and the final sale and dispersal of his goods amidst a growing cloud of fantastic legendry. It was, altogether, just such a story as one would laugh at if possessed of no particular key; yet to me, remembering those dream messages and having Robert Grandison's corroboration before me, it formed a positive confirmation of all the bewildering marvels that had been unfolded.

But as I have said, there is still another line of rather positive evidence—of a very different character—at my disposal. Two days after his release, as Robert, greatly improved in strength and appearance, was placing a log on my living-room fire, I noticed a certain awkwardness in his motions and was struck by a persistent idea. Summoning him to my desk I suddenly asked him to pick up an ink-stand—and was scarcely surprised to note that, despite lifelong right-handedness, he obeyed unconsciously with his left hand. Without alarming him, I then asked that he unbutton his coat and let me listen to his cardiac action. What I found upon placing my ear to his chest—and what I did not tell him for some time afterward—was that *his heart was beating on his right side*.

He had gone into the glass right-handed and with all organs in their normal positions. Now he was left-handed and with organs reversed, and would doubtless continue so for the rest of his life. Clearly, the dimensional transition had been no illusion—for this physical change was tangible and unmistakable. Had there been a natural exit from the glass, Robert would probably have undergone a thorough re-reversal and emerged in perfect normality—as indeed the colour-scheme of his body and clothing did emerge. The forcible nature of his release, however, undoubtedly set something awry; so that dimensions no longer had a chance to right themselves as chromatic wave-frequencies still did.

I had not merely *opened* Holm's trap; I had *destroyed* it; and at the particular stage of destruction marked by Robert's escape some of the reversing properties had perished. It is significant that in escaping Robert had felt no pain comparable to that experienced in entering. Had the destruction been still more sudden, I shiver to think of the monstrosities of colour the boy would always have been forced to bear. I may add that after discovering Robert's reversal I examined the rumpled and discarded clothing he had worn in the glass, and found, as I had expected, a complete reversal of pockets, buttons, and all other corresponding details.

At this moment Loki's Glass, just as it fell on my Bokhara rug from the now patched and harmless mirror, weighs down a sheaf of papers on my writing-table here in St. Thomas, venerable capital of the Danish West Indies—now the American Virgin Islands. Various collectors of old Sandwich glass have mistaken it for an odd bit of that early American product—but I privately realize that my paperweight is an antique of far subtler and more palaeologean craftsmanship. Still, I do not disillusion such enthusiasts.

—1934—

THE LIVING EQUATION

Nat Schachner

Not much is known about Nathaniel Schachner (often "Nat" or "Nathan" Schachner, 1895–1955). His early stories were written with fellow pulp author Arthur Leo Zagat (1896–1949), with whom he shared a background in the legal profession. This particular story, first published in the September 1934 issue of Astounding Stories, *is attributed solely to Schachner. Beginning sedately enough, "The Living Equation" quickly develops into something near-apocalyptical. What could be more dangerous than higher mathematical formulae? Why, sentient mathematical formulae, of course.*

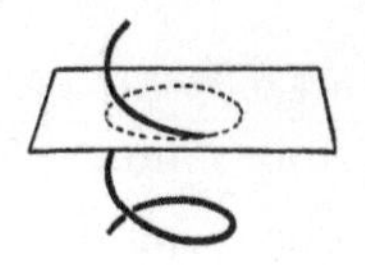

BILL SIKE WAS A BURGLAR; HUGH WILMOT A MATHEMATICIAN. Which was a misfortune. Ever after, Wilmot was to bemoan the fact that their respective rôles had not been reversed. Nor would Sike have raised the slightest objection. Quite the contrary!

Sike, when he picked on Wilmot's small but comfortably furnished home on the outskirts of the great city as a suitable site for the exercise of his talents, had no inkling of what awaited him. If he had, he aggrievedly informed his interlocutor after the event, he would rather have marched to the nearest police station, confessed to the contemplated crime *in futuro*, and accepted with a cheerful and philosophic mind a three-year stretch in the can.

But then, as his guide, mentor, and counsellor, "Louie the Eel," pointed out with much penetrating language, it all boiled down to the fact that he, Bill Sike, was a disgrace to the profession, a blundering idiot, and a stumbling amateur. Catch a skilful craftsman like Louie arousing a household in the course of his nocturnal business, or falling all over a lousy infernal machine in a mad attempt at a get-away!

So the world was thrown into a state of convulsion from which it has not yet extricated itself, without at the same time having any clear idea of the exact mechanism of the debacle. Bill, for particular reasons associated with the laws on the statute books governing breaking and entering, maintained a discreet silence, except to the aforesaid Louie the Eel, and to Hugh Wilmot, who was able to exercise a certain compulsion.

But Bill, as has been stated, was not a mathematician. He could not remember, under Wilmot's most stringent cross-examination, just what it was he had done, and the almost supernatural phenomena of which he was an unwilling witness and sole human participant had left him reduced to a state of incoherent gibberings. As for explanations, alas—

Wilmot knew the explanations. He had been in the habit, when the hour was late and the wine good, of expatiating to his friend, Arnold Polger, certain fanciful theories which Polger, a lawyer and an intelligent man to boot, received with a certain open-mindedness because mathematics generally was in all conscience mysterious enough.

But it was the method, the accidental concatenation of levers and light impulses and whirling disks which set the machine into strange motion, that excited Wilmot to frenzy. Bill squirmed and growled and grovelled, and swore that he had been too busy with panicky flight before, and too terrified after, to know what he had done. The machine, of course, was wrecked, together with everything else.

Wilmot, being a mathematician, calculated the possible permutations and combinations of all the levers and light cells and disks he had incorporated into the damned thing to be a staggering figure with enough noughts on the end of it to take five thousand men five thousand years to exhaust every possible combination. That is, even if he could reproduce exactly what had taken him seven years of ceaseless toil to construct.

So it was that Wilmot tore his hair and fumed helplessly and swore at an unkind fate that had not destined him to be a burglar from his cradle, so that on that eventful night of September 16th, he could have burglarized his own home and crashed into the machine himself.

This is how it all came about.

Arnold Polger followed his host patiently through the library and living room into the rather large chamber that had been built to accommodate week-end guests. He was in an expansive mood. Sally, the cook, had thoughtfully provided all his favourite dishes, and the cigar in his mouth had an excellent aroma.

He liked visiting his friend, Hugh Wilmot. Law was an exacting profession, clients sometimes bores and judges a curse, so it was a relief to steep himself in the rarefied, but extremely exciting, atmosphere that eternally surrounded the mathematician. It sharpened his brains, honed them to razor edge, even though he could not often quite follow the glittering dance of equations.

Wilmot flung open the door. "This," he announced with a certain pride, "is what I wanted you to look at."

Polger surveyed the machine with considerable interest, but no understanding. It filled half the chamber. An intricate array of gears meshed on gears; innumerable levers connected with turning wheels underneath the supporting stanchions; and thick cables ran to an auxiliary motor. Photo-electric cells cast a strange bluish glare on every disk and gear of the complex mechanism. And midway between machine and ceiling, motionless, suspended from no discoverable source, was a smooth, round, transparent globe, a hollow, thin-walled bubble.

"Well?" he said at last. "What is it?"

"An equation producer. I've worked on it for seven years, and I think I've got somewhere."

"Oh!" Polger said disappointedly. He had been intimate with Wilmot long enough to understand mathematical phraseology.

"Something like the 'brain' machine at Massachusetts Tech. for solving differential equations, or the tide calculator at Washington, eh?"

He was really sorry. After all, his friend was brilliant, and he had received in the course of years many cryptic hints of this ceaseless toil of his. He had expected something more.

Wilmot glared at him reproachfully. "I said equation *producer*," he said, "not equation calculator. I choose my words carefully. Those, clever as they are, are mere glorified adding machines. They solve the differential problems that are fed into them, problems for which any mathematician of modest attainments could get the answers with pencil and paper and a few hours' work.

"This is something different. It starts where the others leave off; it is a mathematician on its own hook."

Polger puffed comfortably on his cigar. "Suppose you explain," he suggested.

Wilmot gazed raptly at his creation. "You notice the hundreds of levers and disks?"

"Yes."

"Every one of them, when set, represents a definite component in an equation."

"Exactly the same as the Mass. Tech. 'brain'," Polger pointed out.

"In a way," Wilmot admitted. "But with a tremendous difference. These deal with vectors, directional mathematics, and tensors, the mathematics of strain, the most advanced and intricate of all forms of modern analysis. Without these twin flights of man's genius, we should be helpless to explain the exact inner workings of the universe." His eyes held a far-away look.

"Sometimes I think that these symbols, the little scratches we make to represent vectors and tensors, *are* the universe, the structure itself, and that what we see, the physical, the mechanical, are mere outward clothings of the eternal mathematical thought. We—"

"In other words, your machine is an improvement on the others," Polger interrupted hastily. He had heard Wilmot's peculiar theory before, and it still didn't quite make sense to his practical mind.

Wilmot smiled grimly. "Not at all! My machine will, I hope, build from where the best of us leave off. In other words, it will take our most complicated tensors and vectors and proceed as though it were itself a supermathematician. It will construct of its own accord new problems, new equations, beyond anything I, or any one else, have been able to accomplish so far. It may even," and Polger was surprised at the light in his friend's eyes, "create an entirely new order of mathematics, something as far beyond tensors as they are beyond the multiplication table."

Polger's cigar had gone out. He forgot to relight it. "But then you have created a real thinking machine," he said slowly.

Wilmot nodded. "Exactly!"

The lawyer was bewildered. He looked again at the complex array of steel parts and photo-electric cells. It was exactly the same as before—incomprehensible.

"But how? What force—"

"Light, electricity, energy waves," Wilmot responded promptly. "It all goes back to my theory, the one you listen to without grasping at more than the hem of my meaning. Mathematics *is* real. The higher equations, educed solely from the inner consciousness of our greatest minds, without former roots or counterparts in the world of what to us is physical reality, have, for all their seeming violation of common sense and things as we see them, satisfactorily explained the constitution, the very essence of the universe."

"I still don't—"

Wilmot disregarded the interruption. "I felt therefore that these symbols we evolved were as much existent and endowed with a life of their own as, let us say, the Earth and stars and the atoms whose actions they explained and predicted. Suppose, I thought, we could actually clothe these abstract symbols with physical reality, make them manifest to our senses, would they not interact, produce new forms, even as hydrogen and oxygen in close contact will form a new substance—water."

Polger lighted his dead cigar with a shaky hand. Had his brilliant friend gone mad from too much pondering?

"I am not crazy," Wilmot said quietly. "I succeeded. Each disk can be moved to a position corresponding to a directional vector. Each lever, depending on the amount of thrust, warps the flat steel and creates a condition of stress and strain corresponding to a tensor. A steady beam of light, playing upon the warped, directed disks, suffers certain minute changes. This accords with the theory of relativity. The rays are then reflected to the photo-electric cells, each contacting a single disk.

"Electrical currents are set up in the cells, and these in turn are focused in beams within that globe you see suspended in mid-air."

Polger, his head spinning a bit, craned his neck upward again. "I was wondering about that," he said.

"That," said Wilmot, "is my greatest creation. By a series of strategically placed magnets, thermostats, electrical, and other controls, I have eliminated or neutralized every possible force that might have acted upon that quartzite globe. To all intents and purposes the vacuum within is a section of empty space, quiescent in a universe of its own. As far as it is concerned, the rest of the world of matter and force has no meaning, no existence."

Polger gasped and looked a bit more respectfully at the motionless orb. "And what may the purpose of all this be?" he asked faintly.

"It permits the electrical impulses, each slightly varied by its vector and tensor component, to interact without the disturbance of any outside force. Nowhere else in all space does such a condition obtain."

"I begin to see," Polger said thoughtfully. "And you think that the mathematical waves, so to speak, finding themselves free of all interference, will combine to form new equations, which in turn will fall into new combinations—new to humanly acquired mathematics?"

"Exactly!" Wilmot grinned delightedly. "And I've evolved a method for imprinting the new series on revolving drums from which it will be possible to effect intelligible translations."

Polger felt himself in the presence of great forces. He was awed. Sudden desire swept through him. "Try it out now, Hugh," he said quickly. "If it works—"

Wilmot smiled and shook his head. "It isn't as easy as all that," he said. "I haven't completed the series of tensors I want to feed into the machine. I want it to be the highest flight yet of the human mind, the next step after Einstein's world equation of ten variables. Tomorrow it will be finished. And tomorrow night— But we'll sleep on it, old friend."

A half hour later the house was dark and still. Wilmot, strangely enough, snored soundly, while it was Polger who tossed and stared wide-eyed at the blackness of the night outside his window. Sally, the cook, in her room in the attic, dreamed, as she did every night, of the handsome prince who came to solace her subconscious for her dry and withered spinsterhood.

Bill Sike moved silently to the rear of the house. Thick clouds obscured both moon and stars, and the house itself was but a vague shadow. A perfect night for a burglar!

He flicked his pencil flash cautiously at the low-lying kitchen window. Louie the Eel had taught him that early in the game. "Most people," remarked Louie sagely, "is dumb. They lock their doors with fancy locks an' patent chains; then they go ahead an' leave a window wide open."

Bill chuckled. Smart bird, Louie! For, sure enough, the upper sash gaped invitingly. Within seconds he was inside, feeling his way carefully through blackness. The next room was the dining room. He cast a small questing pencil of light over its contents. An old-fashioned sideboard of rubbed mahogany took up almost the entire opposite wall. Bill grunted his satisfaction. Sideboards meant silverware, and silverware was convertible into cash. Louie the Eel took care of that. He shifted his sack to his other hand and advanced.

In so doing, he brought his shin in sharp contact with the edge of a chair. The chair moved violently against an end table, which tottered and sent a lamp crashing. Bill himself howled with anguish at the pain to his shin. He had not learned the art of repression.

Arnold Polger, still awake—his restless thoughts had travelled from his friend's machine to more important things, to wit, whether the bill he had sent to a certain client had not been underestimated—sat bolt upright. He shouted fatuously: "Who's there?"

Sally, with the salt kisses of her dream prince wet upon her lips, thudded back to reality and screamed. Wilmot shot from healthy sleep to instant awareness. His first thought was of his precious machine. He left his bed in a single bound, and his bare feet made quick patter on the carpeted stairs. Polger groaned and followed, stubbing his toes in the blackness.

Sike was aghast. His first solo burglary had ended in disaster. His shin throbbed and his flash had flown from his hands. It lay at the farther end of the room, making a pool of radiance. He glanced wildly around.

Feet and accompanying voices were catapulting down the rear stairs, cutting off flight through the kitchen, the way he had come. A door loomed dimly to his left. He darted through it, cursing.

Some one yelled "Stop!" behind him. Panic gripped his vitals; he lost his head. He smashed straight on, missing the door to the outside and safety, smacked his shoulder against the opening to the spare chamber, spun around, howled, and went headlong into the darkness within.

Cold metal caught at his feet, tripped him. He thrust his hands out blindly to save himself, and sprawled, flat, outstretched, heavy, against serried banks of levers, buttons, and disks.

At once there was a whir. Things moved and slithered beneath him. A humming sound smote at his ears. Outside there was more sound—the noise of an aroused household.

He scrambled off the spinning contraption, whining in self-pity. Then he screamed in good earnest. His head was back and his eyes glared. He forgot the inglorious end of his venture, the looming hoosegow.

Up above him, suspended in mid-air, was a huge shining globe. The light within it was ghostly, shimmering. It looked to his affrighted imagination like a bodiless head come to plague him for his sins.

Even as he stared, paralysed into stone, mouth still open from his scream, the globe swirled with strange green currents which grew thicker and thicker until they seemed liquid, oily.

Fantastic configurations appeared and disappeared within its depths, succeeded each other in rapid succession. Cubes, rhomboids, cones, paraboloids, then shapes and forms beyond all human or physical conceiving—fourth dimensional, fifth, multidimensional, swirled and melted with the rapidity of their contriving. Higher and higher they built up, pure syrupy vibratory essences, until the whole mass seemed to explode of its own complexity.

The globe expanded into an unknown dimension and disappeared suddenly. Waves of green luminescence beat outward, surrounded the burglar, pierced through the confining walls of the room as though they were so much glass, went on and on with unimaginable rapidity, beyond the speed of light, past the substantial stone of the house, into the night, engulfing, swallowing up, fields and rivers and towns and the solid earth itself; New York, St. Louis, San Francisco, swift-cleaving liners in mid-Atlantic, soft green isles in the South Pacific; London, Paris, Moscow, Shanghai, the Gobi, Eskimo igloos in frozen Greenland, whales spouting in Ross Sea.

Out into space the radiance went, impacted on the Moon, pierced it through and through with green fires, expanded outward with ever-increasing velocity. Within seconds Mars, Jupiter, Pluto were aflame with virescent shine; the Sun shuddered and masked its yellow heat under the tremendous influence.

And still the beating waves did not cease. Space rolled up like a well-scanned parchment before their gigantic strides. Alpha Centauri winked suddenly green; so did Sirius, Procyon, and far Aldebaran. The Milky Way glowed with the strange gleam and the Galaxy was hurdled. On and on, spanning the nebulae, racing through island universes, past Andromeda, until the very outposts of the scheme of things entire were far behind. Out into the emptiness, where matter did not exist, where space itself was a figment, where the waves of mathematical probability were the only reality in the inchoate, unstirring dreams of creation.

And the heart of the universe, the core of all things, was in a room in Wilmot's house.

A ball of liquid green light hung motionless where the globe had been. Within its strange consistency was a paradox. Featureless forms, passive activity, abstractions made visible, mathematical equations of dizzying complexity clothed in ceaseless vibration.

Bill Sike moaned and remained rooted to the floor. Terror robbed him of speech, of movement, of any human faculty of comprehension. Outside, Hugh Wilmot, pyjama-clad, hurled himself toward the open doorway and crashed into an invisible wall that sent him back, battered and reeling.

Arnold Polger, barefooted, came panting up. He stopped abruptly, stared. "What the devil's happened?" he gasped.

Wilmot did not answer. He crouched like a linesman, smashed forward again. He brought up short, his shoulder compressed, and once more he fell back, face twisted in agony, frenzied with something beyond agony. Yet no obstruction showed.

Polger plucked at him with restraining hand as, insanely, he bent to resume his futile charge. "Don't be an ass," he said sharply. "You'll kill yourself."

Wilmot turned his tortured face to him. "Don't you see?" he

almost sobbed. "The machine—it's worked! Some one—that man in there—started it—somehow—"

Polger peered through the luminescence that bathed him, together with the universe, and saw the fantastic ball within, the fear-frozen figure of Bill Sike. His legal-trained mind responded immediately. "He's a burglar," he snapped. "I'll call the police."

"To blazes with the police!" Wilmot shouted. "I'd give a million dollars to be in there, in his place. Something has happened that lay beyond my wildest dreams."

"What?"

Sally, stringy grey hair in iron curlers, voluminously wrapped in a flannel nightgown, stumbled down the stairs. The queer green radiance was unmerciful to her sallow, dried-out features. She saw the globe of light, saw Bill Sike, who, released from his paralysis, made a wild dash for the door, to be smashed back from nothingness and converted into a gibbering, moaning wretch. She screamed shrilly.

"Shut up, Sally!" Wilmot snapped.

But Sally was beyond hearing. She was already out of the house door, into the green luminescent night, running, sobbing, shrieking, cutting her bare feet on stubbly grass, crying doom to all the world. Nor was she alone on her wild flight that night.

Wilmot caught his breath. "It's mad, *mad!* The equations have formed in the globe; they have spawned and combined and propagated new equations—I catch faint semblances of mathematical abstractions clothed in being within; but they've done more than that. They're affecting the universe; changing its laws maybe."

He glared at Polger in a sort of futile frenzy. His voice rose almost to a scream. "They *are* the universe! You, I, are mere illusions, manifestations of their eternal configurations." He gripped his friend's arm with fierce strength. "Good Heaven, man! Do you realize what it may mean? Suppose they spawn a new type of mathematics, non-Euclidean, non-Riemannian, non-tensor, non-vector, something that does not fit in with the present laws on which our universe is constructed."

"Well, suppose they do?" Polger said a bit impatiently. So far, it was a mere matter of light phenomena due to that peculiar globe which, strangely enough, had disappeared. He was more interested in catching the burglar. The man might come out of his grovelling fit any moment and make a break for it.

"You still don't understand!" cried Wilmot. "Our universe would vanish like smoke into the limbo of forgotten things together with the underlying equations which were its basis. Another universe would take its place, consonant with the new mathematics. That light is spreading, Heaven only knows how far. It's alive, sentient, instinct with the newborn equations. It will devour, destroy, the old, the accustomed."

Polger stared blankly. His practical mind fumbled at the implications. "And what will happen to us?" he asked feebly.

Wilmot laughed bitterly. "Us?" he echoed. "We'll be dissolved with the rest of the illusions. There'll be no place for our symbols in the higher mathematics. If only I could get in there, to stop the machine, to see how it is set."

He shouted, hoping to break through the wall of force with his voice, to reach the grovelling burglar, to make him understand what to do. But Sike rolled on the floor in a veritable fit, nor could he have heard in any event. The force bubble that had been thrown around the room was impenetrable to any agency known to man.

*

Polger still did not believe. He shifted uneasily from one bare foot to another, acutely conscious of his state of undress, of the fact that the uncanny green light was playing nasty tricks with their complexions. He wanted to call the police, the fire department. Those very practical embodiments of law and order in a universe of law and order would know just what to do.

Wilmot's cry brought his head jerking up. "It's come! The new order of equations. They've just spawned. Our universe is done for. Good-by, old friend! If only I could understand, get in there—"

His voice trailed off; he fell unconscious.

Polger lasted a while longer. Perhaps his physical configurations conformed a bit closer to the new laws. Bill Sike lay inside, mouth agape, foam frozen on his lips, eyes protruding, to all appearances dead.

Polger saw the uncircumscribed ball of green liquid flame change slowly to a pure supernal orange of dazzling hue. Figures, vibrations, swarmed in ceaseless whirling within. Then suddenly, there was a soundless explosion, the strange component abstractions fell into a pattern—a pattern into the depths of which Polger's mind penetrated for one instant of breathless, awful comprehension.

For that one instant he was in tune with the new universe. Then, as the ball expanded and rushed to engulf him, he, too, fell unconscious. Never after was he able to explain what it was he had seen, what for that one dazzling second he had comprehended. Wilmot swore and fumed over him, even as he did over Bill Sike, without result.

Polger shook his head feebly and gabbled about forms that had gone beyond all dimensions, that possessed no dimension at all, forms for which time had no meaning or existence, and which—most stupendous of all—were pure thought, pure consciousness, pure life, stripped of all physical excrescences.

Wilmot, the one man who could have explained, who could have duplicated the phenomena, had not observed. It was given to a burglar to initiate and a lawyer to witness the finale, and none other. For which the universe as we know it, to the farthest nebula, should give heartfelt thanks. It was close enough to destruction as it was, and Wilmot's caution could be ill relied on in the presence of greater and vaster scientific experiments.

The night mail was flying rather high over the Alleghanies. The passenger, a radio star in a hurry to get to a big movie contract in Hollywood, leaned forward toward the cockpit.

"Everything seems lighted up with green lights," he said conversationally. "What's up—a celebration?"

The pilot heard him perfectly. For the first time he realized what was wrong, what had been bothering his subconscious for the past few minutes without his being able to lay a finger on it.

The deathly silence; the absence of all sound. The roar of the propeller was muted to nothingness. They seemed to be swimming in a motionless sea of green flame. For a wild moment he thought they were falling, yet his instruments showed motion, and the body throbbed from a noiseless motor. The lights, too, bothered him.

He swung half around, anxiously. "I don't know just what—"

Orange blasts enveloped them. The next instant both men felt violently compressed into themselves. It was the last sensation they ever had. The plane and its occupants vanished; in their place a tiny, microscopic globule—electrons and protons compacted into a unitary mass—fell with terrific velocity toward the earth, to bore of its own incredible weight a mile below the surface.

The great liner, *Ladonia*, nearing the coast of Europe, was ablaze with lights and merriment. It was the last night on board, and a fancy-dress ball was in progress. The saxophones moaned and the violins tossed off spangles of glittering notes. Fairy princes danced with Carmencitas and whispered words into small ears that brought quick laughter. The huge punch bowl ebbed and filled with magic rapidity.

The bar was crowded with men, flushed of face, beefy, righteously scornful of being togged out like Boy Scouts. Couples snuggled on steamer chairs near the lifeboats, oblivious to dance and bar and sky. A woman strode purposefully along the deck, peering into the faces of the bemused couples. Their faces were green with strange illumination, yet they did not notice.

The woman, hatchet-face grim with repressed fury, pursued her tour of investigation. In the shadow of a lifeboat she spied two dim figures locked in each other's arms. Lips were pressed tightly together.

She bore down on them like a destroying demon.

The man heard, raised startled eyes. "Maria!" he gasped. Without another word he tore out of the arms of his lovely companion, and fled down the deck as if all the devils of hell were after him.

The woman turned on the girl: "You dirty tramp, you—"

The captain stood on the bridge, legs astraddle. He stared into the night. "I've never seen an aurora like this before," he told the first officer.

"No more have I," said his subordinate. "It came up like a blanket. Look at the moon, too, and the stars. They're all green."

The telephone rang. The steersman said excitedly, "The compass has gone haywire, sir. And the ship doesn't respond to the wheel."

Another telephone jangled—the engine room. The shocked voice of the engineer. "Steam's gone, sir. And the fire's out; the coal lies dead and cold. Tried to light a match, sir, and the blamed thing—"

The universe turned orange.

The liner *Ladonia*, tilted, shuddered, and rose straight into the air. A huge waterspout followed her.

The captain sprang toward the pilot house. The ship tilted more, and he sprawled out into the open; a whirling tiny figure that rushed up and up, keeping even pace with the soaring liner.

The fleeing man, avoiding the wrath of an aroused wife, stumbled and catapulted over the rail. He, too, became a satellite of the ship.

The fancy-dress ball was a confusion of struggling, screaming, and floating creatures. The ship accelerated its mad upward flight; ice formed a quick curtain; the cold of the stratosphere bit into the thinly clad bodies of the passengers. Their screams grew more and more feeble; blue congealment stilled their struggles.

Then the air rushed out into space with a great *swoosh*; the limits of the atmosphere had been reached. The frozen bodies developed steely hardness; those few who were still alive literally exploded from release of pressure.

Higher and higher, faster and faster, the great ship fled, followed by masses of ocean turned into glittering ice, followed by the dead captain and the man whose wife had caught him in *flagrante delicto*. A huge cosmic morgue, with thousands of dead bodies floating and bumping within its vast interior, the liner sped upward and outward, past the Moon, past Mars, out into the region of the asteroids.

By that time the universe had settled back to normal, and the *Ladonia* became a new asteroid, swinging around the Sun in a majestic period of five years and thirty-four days.

The laws of gravitation had been reversed for the particular sector of the Atlantic Ocean in which the liner had unfortunately been.

For the new mathematical order of things proved positively freakish. Vast portions of the universe were left wholly untouched; others had only certain limited modifications in the laws governing being; circumscribed areas received the full impact and vanished or were irreparably changed.

Wilmot tried to explain it afterward. "You see," he told Polger, "my machine was not quite powerful enough. The superequations filled with the universe, it is true, but spottily. Had their influence been evenly propagated, nothing would have remained, not even space and time. Then again, the equations themselves realized before it was too late—"

The New York telephone exchange was extremely busy, even though it was past midnight. Girls said monotonously: "Number, please! Thank you, I'm ringing them. Sorry, sir, the line is busy. No, madam; I don't know what is happening."

All New York had awakened at the curious green glare that shrouded the city, blotted the heavens, masked the normal yellow glow of millions of lights. All were possessed with the same idea. This was the weather bureau's job, for which good citizens paid taxes. Ask them about it.

James B. Wales, night editor of the *Clarion*, jiggled his phone furiously. "Listen, sister," he said hoarsely, "I got to get the bureau."

"Sorry, sir, line's still busy."

"T'hell with that. I'm the *Clarion*. Cut 'em all off and give me a wire. Be a good kid; I'll take you out to dinner, *and* a show."

The phone went dead. He swore and pushed black streaming hair out of his eyes.

"The dirty so and so!" he said in low earnest accents.

But the girl was not to blame.

For just then the orange wave lashed out.

Where the gigantic setback cliff of steel and tan stone had stood was a gaping hole. The Telephone Building was gone, vanished forever, never to return. Some topsy-turvy shifting of a mathematical formula in its vicinity had snatched it, thrown it headlong into a superdimension. It might have been right where it had always been, but our normal dimensions did not contain it.

A year later, when all hope of its return was abandoned, a new and larger building was erected on the spot. Yet Wilmot always had an uncomfortable feeling that the ghost, the superdimensional ghost of the kidnapped structure, glared forlornly at and through him every time he entered its precincts.

For by some strange unaccountable freak, telephone communication with the world as it is had not been lost. Engineers traced it eventually to a wire that seemed to taper into nothingness. A guard was placed around the wire to prevent accidents. The guard was maintained for at least a month after the final message had come through.

Nellie McCafferty was the heroine of the tragic misadventure, the girl whom Jimmy Wales had called unmentionable names. He took every one of them back, blazoned her name all over his headlines, raised a whacking big fund for the relief of the parents whom she had supported out of her scanty salary.

And no wonder!

For through her he achieved the biggest scoop in the history of the newspaper business. Imagine talking directly with a girl in another dimension, getting her sensations, the sensations of a thousand other men and women in that crazy place; people, who to all intents and purposes, were dead, vanished.

It was his connection, cut off by the orange flash, which still functioned.

For days she stuck to her self-appointed task. They were, she told Jimmy, in a world of utter blankness. No light, no sound, no depth, or height, or width; nothing but a flat impenetrable greyness. Of course, said Wilmot, when he heard of it. They were still three-dimensional people in a three-dimensional building. The new dimensions could mean nothing to them.

At first the thousand marooned humans called frantically for rescue. They did not understand, did not realize why, when they tried to leave through open doors and windows, there was nothing for them to step into. It was a curious sensation, which, in spite of faithful efforts to reproduce for the benefit of the people on Earth, held no meaning.

Then awareness came to them, and with it fright. Rescue from their predicament was impossible. Though doubtless as near to their own kind as interpenetrated bodies could be, they might as well have been out on Orion's belt.

For weeks the world hung with fascinated horror on the slowly unfolding tragedy. Only a thousand lives were involved, as against the millions who had died all over the Earth, and the possible countless billions of forms who went crashing in debacle throughout the universe. But here was novelty, gigantic drama, human interest, everything that went to make up a perfect play. Ghosts from beyond our time and space talking with us daily!

And the *Clarion* grew fabulously in circulation, and Jimmy Wales was content.

Nellie McCafferty told in plain, unadorned words the story of their growing fear, the trapped feeling. Then came hunger; thirst a little later. The supplies of the company's cafeteria were scanty and had to be rationed. Then came a worse difficulty; the air in the building was giving out, growing foul and stale with overuse.

Day by day she communicated; slowly, more painfully each day. Then finally, the phone rang.

Wales lifted it; there was a faint whisper, a gasp. "Yes, yes, Nellie!" he half shouted. He swears, and so that night's edition had it, that he heard her say: "Good-by, people of Earth. I am the last alive."

Then, and this was indubitable fact, the connection broke, irrevocably. The superdimension had lost contact.

To Jimmy Wales' eternal credit be it stated that he had to wink his eyes rather violently for a moment; then once more he was the hard-boiled editor.

"Copy desk; rewrite man, boy, press-room!" he bellowed. "Stop everything, take this for the final. Wow, what a story!"

The teeming province of Shantung in China disappeared, dissolved as though it had never been; land, people, animals, rivers, thereby bringing to an abrupt end the plans of outside nations to gobble up this delectable bit of territory.

In its place was a hole, a gigantic incredible hole extending over thousands of square miles of area and descending in a smooth bottomless pit a trifle less than a thousand miles into the interior of the Earth.

Geologists and scientists of all nations, with scarcely a thought for the millions of unfortunates whose mathematical laws had vanished and with them their own existence, descended with loud glee upon this unparalleled opportunity to see the very bowels of the Earth bare.

For over a hundred years they were busy, inventing machines to take them down where no man had been before, studying strata, examining specimens, wrangling, concocting new theories to fit new facts, living, growing old, bearing a new crop of scientists to take their places, all in an unprecedented state of excitement.

Luckily sparsely inhabited, North Australia stood up on end for hundreds of miles, to remain a fabulous mountain visible for thousands of miles. The Pacific rushed into the depression and formed a new ocean. As a result, the general level of the Earth's seas was lowered, and Atlantis was rediscovered. It proved to be in the shallows of the Caribbean Sea. Stately buildings, unimaginable artifices of ancient people, rose, covered with slime and mud.

But there was another more serious effect of this upending, coupled as it was with certain other changes in the structure of the solar system. The delicate balance had been destroyed. The Earth's year was cut to three hundred and thirty-six days, seven hours, two minutes, and prominent scientists issued grave warnings. At the present rate, all other things considered, the year would become shorter and shorter, and within fifty million years, Earth and all its works would fall into the maw of the Sun.

At first, however, the news had been very much more alarming. The Sun and planets of the solar system were functioning, with slight modifications, according to former schedule. But the rest of the universe seemed to have gone haywire. The background of stars went round and round in streaking circles of flame. It was impossible to register the number of revolutions per day, so fast did they swing from pole to pole. Not only that, but the hitherto fixed stars were actually moving across the face of the heavens, visibly, shifting whole minutes of arc from day to day.

Astronomers turned grey overnight. Rough calculations showed speeds that were absolutely unbelievable—trillions of miles per second. It was impossible; it was mad; it was insane! Wilmot finally hit upon the explanation.

Time had changed for the solar system. A new system of coordinate functions had displaced the old, while the rest of the universe had remained unaffected.

Time had speeded up in the region of the Sun and its planets. A thousand years of sidereal time had been compressed, for us, into a single instant of existence. The rest of the universe was growing old while we still remained infants, so to speak.

This strange state of affairs made no appreciable difference in our private concerns, inasmuch as, aside from the clock of the stars, life seemed to go at the same pace as before. But it played the very devil with sidereal astronomy. All calculations, all former data, had to be revised.

Nor were these the only profound disruptions due to the release of sets of formulae and equations that had no former counterparts in the universe. Stars blanked out completely; others were discovered after much painful searching in an entirely different part of the heavens. Instead of an expanding space, the nebulae seemed simultaneously to have made up their minds to return to the primeval gigantic atom from which they had originally emerged.

New life forms appeared suddenly on Earth, strange sentient creatures whose physical basis was silicon, who neither breathed nor ate nor seemed to move. Nor was this inexplicable. The mathematics of evolutionary processes had shifted, too.

Fortunately they all soon died.

The maddest freak of all, however, was the simultaneous appearance of formerly respectable material things in different parts of the world. The Empire State Building, for instance, found itself in the middle of the Arabian Desert. Travellers, whose reputation for unswerving honesty and sobriety was unimpeachable, reported it there, solitary and somewhat ashamed. Yet the Empire State Building still occupied its accustomed quarters at the corner of Thirty-fourth Street and Fifth Avenue. Mount Everest, while still a fixture in far-off Tibet, also achieved a second domicile not far from Washington, to the delight of intrepid mountain climbers who had not the wherewithal to make the expensive and arduous journey to the Himalayas.

Not so pleasant was the predicament of a poor devil of a bank clerk who found himself suddenly three people, wandering the streets of the same city, meeting himself over and over again, to his own vast confusion and to the utter bewilderment of a wife and four children who had not followed him in his metamorphosis. Quite naturally she raised a to-do about such goings on, and quite as naturally each of the three simulacra objected to further support of a family, who, each argued, refused whole time conjugal and filial duties.

At last the matter reached the courts and became a *cause célèbre*. The case dragged for seventy-five years. When a decision was finally reached, all of the original participants were dead, and the learned jurist who wrote the closing opinion laboured under the reasonable misapprehension that it had something to do with a remission of taxes on a non-existent plot of land.

Wilmot slowly became aware of his surroundings. He was flat on his back and his side hurt. Polger lay near him, breathing stertorously. A bit farther off, eyes closed, and moaning through clenched teeth, was the burglar, Bill Sike.

Where was he? What had happened? He opened his eyes. The green light had gone; in its place was darkness streaked by curious wheels of fire. Underneath him, too, was damp strangeness. He felt weakly with his hands.

Earth and grass and small stones, wet with dew. He was out on the lawn, in the open. He sprang to his feet, panicky. How had he been carried out of the house? What about his precious machine—the mathematical life to which it had given birth?

He strained feverishly. It was his own lawn, without doubt. There were the two beech trees in which he took so much pride; to the left, as ever, nestled the little rock pool.

But the house—it was gone! A depression showed where it had stood.

Polger staggered to his feet, eyes bloodshot, staring. Sike instinctively started to creep away on hands and knees.

"No, you don't," Wilmot said harshly, and grabbed him by the collar.

Bill whined. "Please let me go, mister. I got five orphaned

children, an' I only did it to get 'em some grub. Please don't send me t' jail."

Wilmot laughed wildly. "Jail? There *are* no jails. I want you to tell me exactly what you did." He shook the shivering burglar. "Do you hear?"

But as has already been stated, Bill's information was not illuminating.

Morning came on a stricken universe. Wilmot's house was definitely gone. With it the machine, the cause of the tragic adventure, and also the mathematical globe of equations.

It was decided to keep quiet about the machine.

But Wilmot and Polger discussed interminably in the privacy of their own quarters.

"What," asked the lawyer for the hundredth time, "happened to the globe of equations?"

Finally Wilmot said hesitantly: "It may only have been a dream, but while I was unconscious, it seemed to me that I was aware of certain things happening in the globe. The configurations, the pure inherent thoughts, were dissatisfied. They had conformed parts of this universe to their own laws and theorems, but not all. The old laws were in the main too powerful. There was tremendous conflict.

"It was therefore decided to build a universe of their own, completely outside our time and space, where the outward habiliments would be the harmonious counterparts of their equational desires."

Polger pondered a moment. He rose. "Thank God!" he said with due deliberation.

—1936—

INFINITY ZERO

Donald Wandrei

Another end-of-days story from the pulps, this time by Donald Wandrei (1908–1987). Born in Saint Paul, Minnesota, Wandrei's writing career began in 1928 with a book of poems. By the late 1930s he had authored over two dozen stories for the pulp magazines. Perhaps Wandrei's most significant role in the history of the weird, however, is the part he played in the creation of Arkham House publishers, which he co-founded in 1939 with Lovecraft devotee August Derleth. Initially formed as a means to keep Lovecraft's work in circulation, Arkham House quickly garnered a reputation for producing fine hard-back editions of weird and supernatural fiction, early editions of which are still highly sought-after. "Infinity Zero" made its debut in Astounding Stories, *October 1936. Some impressive description and an inexorable climax make it one of the more existentially terrifying stories of its kind.*

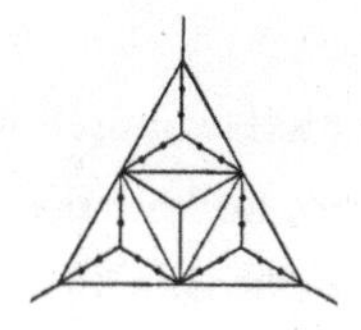

CONWAY, A SPOT NEWS PHOTOGRAPHER FOR THE *Herald*, WAS staring moodily out of a grimy window when the telephone rang. He paid no attention to it because it wasn't his job to answer. Anyway, the telephone rang all day and all night long.

Across the Hudson rose the dark towers of Manhattan, limned against the faint glow of the sky. No lights shone in those giant structures, for the second year of the war had made cities the world over reluctant to use electricity after nightfall. Bombing raids of the German-Italian-Asiatic League had already ripped gashes across Boston, Washington, and other coastal cities. The government operated from the interior. The American-Soviet-Anglican group—a union of strangely assorted allies—had laid waste foreign capitals—Tokyo, Berlin, Rome.

A quick thrust and victory within a month had been forecast at the beginning of the war; but, as in the old World War, the opening thrust failed, and the war settled down to a steady, remorseless virulence that marked the beginning of the long-promised decay of civilization. Oceanic shipping had virtually ceased. Disrupted transportation, commerce, and communication upset the whole intricate balance of society, threw local communities back upon their own resources. There were no television broadcasts after nightfall.

Conway was conscious of some one answering the phone. He continued to stare at the lofty spires across the river. When would it end, this senseless slaughter that mowed down millions by poison gas, explosives, plagues, famine, electronic radiations, robot planes and rockets laden with death? He himself would be out there somewhere at the front except for his limp that made the universal draft reject him. It didn't matter. He'd die just as horribly, though a civilian, during one of the periodic raids.

The night editor called, "Conway, cover this."

He turned around with a bored expression and lighted a cigarette while he listened.

"You know the way to the United Chemical Testing Laboratory?"

"Yeah. Thirty miles in and up the hills over roads that weren't even paved before the war started. Heaven knows what they're like now. The sentries and guards won't let any one pass."

"Get through. The place is on fire. Get a picture."

Conway said, "On fire? It can't be. It's all steel and concrete and fireproof. Somebody's kidding you."

"Maybe. It sounds queer enough from the flash—that a bomb struck the laboratory and it went up in flames. You'll have to cover the writing angle, too. There's nobody else to spare and the New York papers will want it if it's good."

"Check."

Conway lugged his camera equipment outside and drove off. He didn't see any drifting lights in the sky or hear the roar of airplane motors, but he watched the glare that became increasingly prominent miles ahead. There were no sentries at the points where he expected to be challenged. He found out why when he parked his car by the gate in the wall that surrounded the U.C.T.L.

The laboratory, before the war and since, had been a cooperative enterprise, a proving ground for the big chemical industries. It was stocked with everything known to chemists, and equipped to test any conceivable new compound, formula, or product. It consisted of only one building, a single storey high, that covered seven or eight acres.

The staff, guards, and sentries stood by helplessly, watching it go up in flames. Conway wondered why they didn't try to stop the conflagration. Probably they couldn't use water because of the deadly supplies within.

No one interfered as he set up his equipment and took pictures. There would no longer be any need for sentries around the property of the United Chemical Testing Laboratory.

Conway's astonishment grew as the fire flamed hotter. It was a sight the like of which he had never seen. Concrete, steel, walls, windows, and other inorganic material burned with a strange fury, an intolerable glare, like the light from a trillion electric lamps, as whitely incandescent as the birth of stars.

"What happened?" he asked a haggard-faced man standing near him.

The stranger answered without turning his head. "We don't know. We were using the cyclotrons and atom-smashers in Lab. 9, and a vat of phosphorus in Lab. 10, when the bomb landed between them. It must have been one of the new bombs, formula unknown. The stratoplane got away. Then this—a total loss. The way some of the men were blown to pieces before my eyes—" He shuddered.

Conway glanced him over, and noted the blackened features, shredded clothing, and wounds that still bled.

He shrugged his shoulders faintly. There was no battlefield in modern war. Civilians, women, children, the old and decrepit, the young, the sick, hospital units, dictators and idiots—it didn't matter which; bullets didn't select their victims. The war spared no one, and nothing.

Conway obtained a few statements from observers and returned to his desk. He wrote a brief story whose novelty saved it from being sandwiched between the war news. For him it was simply part of the day's schedule, and then forgotten.

Four nights later, when he returned from New York City with some photographs of what remained after a rocket bomb had destroyed the Medical Center, the night editor said, "Good stuff, Conway, great for propaganda purposes. But there's a follow-up job—"

"The hell with follow-ups," Conway drawled evenly, a cigarette stub drooping from a corner of his pale lips. "I'm sick. There was an old lady on the operating table when the rocket struck. It killed the surgeon just after he'd cut for her appendix. They found her an hour later, just as dead as if the bomb had knocked her off; but the look on her face—"

"What of it? War is war. It can't be helped nowadays. Death is death, whoever goes.

"Anyway, hop over to the U.C.T.L."

"Again? I covered it a few days ago. Something new in the wind?"

"Yes, something queer; I don't know what. People are evacuating the towns around the place. The press report calls it a flame terror, whatever that may be."

"It's probably the débris smouldering. I thought the owners abandoned the laboratory as junk after the fire?"

"They did. I don't think they know anything about this new development. See what's in it."

⋆

Conway reluctantly drove off on a trip he considered a waste of time. As miles of the northern New Jersey hills slipped behind him, however, he noticed an odd glow in the sky ahead—a glow that reminded him, strangely, of a gigantic worm.

The dirt roads, the absence of traffic, the absolute quiet, and the black masses of vegetation intensified his loneliness. There was a loneliness abroad in the land itself, despite the warmth of the soft summer night. He was at a loss to account for this sense of desolation, until he gradually became aware that not a sign of life showed in the farmhouses he passed at rare intervals.

Conway knew that the residential colony for employees of the U.C.T.L. had been evacuated the day after the fire, when officials decided to abandon the site. But why had tenants of neighbouring farms deserted their lands? Even the total destruction of the laboratory shouldn't have affected the farmers' means of livelihood in any way.

Conway's interest began to mount. Mystery lay ahead. Over him crept an uneasiness, an eerie unrest that he had not felt when the noncombustible materials of the laboratory defied the laws of chemistry. The burning of the laboratory had only superficially impressed him. In the past he had seen electric welding arcs, thermite, annealing ovens. He had assumed that the apparent inflammability of the steel and concrete work of the building was due to some peculiar chemical reaction launched by the explosion of the bomb, and intensified by the contents of the laboratory.

His old 1940-model Streamline crested a hill.

The broad, shallow bowl beyond it had contained, five days ago, the property and buildings of the U.C.T.L. Conway expected to find ruins that perhaps smouldered yet, and which would account for the rumours of sinister forces at work.

He stopped his car, and for a long time stared at the spectacle ahead. There was a basis for the whispers, but a basis of such startling and awesome nature that he sat in darkness until the rim of the moon pushed up on the eastern horizon and pale rays filtered into the basin.

Not that he needed moonlight to see, for above the heart of the valley, above the spot once occupied by the laboratory, hung a pillar of flame, absolutely motionless, approximately forty feet tall, and as grey, as lifeless, as the pallor of death.

The flame possessed other strange features: It did not, so nearly as he could estimate, emit any heat; no smoke rose from its tip; its base failed by four or five feet to touch the ground; it assumed a most extraordinary shape, beginning with a pin point at the bottom, expanding like an inverted cone as it rose, and ending in a perplexing blur that tortured vision. The flame seemed to disintegrate at the blur, to splay in all directions and toward more than three dimensions.

The flame, moreover, did not illuminate the ground underneath it. Conway puzzled over the phenomenon, but could not find a reasonable answer. He clearly defined the leprous pallor of the flame; yet the flame did not cast sufficient light for him to see earth or vegetation.

He watched for at least a half hour before the moon rose. During that interval the flame had grown taller, Conway knew, because he could no longer see a couple of stars above the blur that capped the flame. Its pin-point base, also, was measurably higher off the ground.

As the rays of the moon flooded the bowl, Conway's eyes widened suddenly, and the hairs on the back of his neck tingled. A cold sweat broke over him. Fear seeped into him, and the instinctive horror with which human beings have always regarded supernormal manifestations that neither science nor knowledge could account for.

The flame was poised above a gigantic pit, fully a mile wide, a half mile deep, and in the shape of a mathematically perfect hemisphere. The hemisphere expanded while he looked on, with a definite, and it seemed to him, an accelerating ratio.

Conway couldn't imagine how that colossal crater had come into existence after so short a period. If the chemical reaction that destroyed the laboratory had continued, then surely the edges of the great hole would be incandescently glowing. Yet the earth rock lay exposed in its natural state. Yet something devoured, ate away, consumed the earth rock before his gaze.

The flame grew taller, but its greyness remained—and its curiously lifeless hue, its absence of heat. The flame and the solid rock behaved as though an invisible bubble was expanding, pushing them away, or rather, absorbing the earth and enlarging the flame.

Not a sound disturbed the stillness of the air.

The free destruction of matter releases energy—a transformation always audible to human ears. Conway heard nothing.

He watched the hemisphere swell, the flame mount taller, the blur on top of it broaden. It struck him that the hemisphere in the ground, continued through the space above, would form a true sphere at the top of which rested the pointed base of the pillar of flame.

Nowhere was there sight or sound of any living thing. The insects had gone, the birds had gone, the bats had gone, the rodents and the myriad little flying and walking and burrowing creatures of night, all had gone. Now even the wind was stilled. He listened in vain for the rustle of leaves, the faint whisper of grasses, the thirsting buzz of the mosquito. They all had ceased.

The worst loneliness of his life descended upon him. He felt like the last survivor amid a desolate world.

Conway got his equipment ready and bungled the job of taking photographs. He made every mistake possible, from exposing the same plate twice to forgetting to pull the slide out and failing to open the shutter. It was doubtful if one good picture would result from the dozen plates he used.

He didn't begin to regain his composure until he had turned around and was speeding away from that great, corrosive pit and the unflickering, monstrous, corruptly pallid tongue of incredible fire that towered threatening toward the stars.

Conway raced his decrepit Streamline at its limit on the furious drive back to town. He thought hard all the way. And when he finally squeezed the car against the kerb, his game leg didn't prevent him from taking the steps three at a time.

The night editor glanced at him with an air of bored surprise. "What's the rush? Haven't seen you in such a hurry since the war started. By the way, a flash came through on two big battles raging around Moscow and Los Angeles. Casualties have already reached 400,000. That means another three-deck streamer across the front page, provided—"

"A three-decker, all right, but not that one," Conway cut in. "Yank everything off the front page. Send these plates to the dark room and tell 'em to drop everything else while they rush the prints through. There isn't going to be anything on the front page except the streamer and all the pictures that turn out. If only one is any good, enlarge it so it occupies the whole page. Start my story on Page Two."

The night editor bristled. "Who the devil do you think's giving orders around here?"

Conway flopped in front of a typewriter, ran a sheet of yellow copy paper into position. "I don't know who's giving orders and I don't give a damn. This is a bigger story than the whole war. What's the date today? August 7th? Thanks." He began pounding the keys as fast as his nimble fingers could fly, reeling off page after page.

"Is it a scoop? What did you dig up?" asked his boss, the nose for news triumphing over the dignity of office.

"I haven't got the foggiest notion, and even if it was explained to me in words of one syllable, I still wouldn't know the answer. So I'll just write it the way I saw it and if nobody can make head or tail out of it they can just look at the thing. They won't be able to make any better sense out of it either, but that's their funeral."

"Have you gone out of your mind?"

"Yeah. I'm on the way to the bughouse, and so'll you be after you get a load of this."

Conway typed for a solid hour, without pausing to light the cigarette that his jangled nerves craved. He batted off 4,000 words of copy in that period, four full columns that went straight to the composing room as fast as he finished each page.

The night editor devoured the copy, and personally edited it only for typographical errors. It was a whale of a feature.

As Conway feared, the dripping-wet plates came from the developing tanks ruined for the most part. One good plate, however, would do for the front page, and two inferior pictures could be used inside.

The moment he finished his copy, he fished for a cigarette and looked at the wall clock. It was after one thirty. He picked up a phone and tried repeatedly until he got the number he wanted. The number was listed under the name of "Marlow, C. H., Physicist." Conway was not well acquainted with Marlow, but had interviewed him in the past and met him on other occasions. The write-up had particularly pleased the physicist.

Conway talked a blue streak. He beat down Marlow's reluctance and drowsiness. By the very vehemence of his tone rather than any argument that he was able to advance, he won his purpose.

The night editor noticed him hoofing out, and called, "Stick around, Conway. There'll be other stuff coming up."

"It wouldn't matter. It wouldn't be important if Japan and all of Asia sank in the Pacific. And if you feel like canning me, go ahead. I don't think any of us are going to last long enough for it to make any difference what happens now. Personally, I've always thought you were a pretty decent guy, but you ought to chew cloves."

Conway sat in his car, idling with cigarettes, until a big fast sedan rolled alongside. He stepped into the other machine.

Marlow was a rather chubby man in his forties and given to nervous habits. Beside him sat another man, elderly and lean and white-haired, with a marked serenity of features, as though he had achieved the mythical wisdom of the ages and made his intellectual peace with all things that are. Conway liked him during the brief glimpse he got before climbing into the rear seat among a lot of shiny stuff.

Marlow said, "This is the newspaper reporter—"

"Photographer." Conway closed the door.

"—that I was telling you about. Conway, Professor Daël, the mathematician. He's been my guest for the past few weeks. I took the liberty of inviting him along. I also brought a few basic field instruments in case your report is substantiated. Where to?"

On the way toward the former site of the U.C.T.L.—"site" was good, Conway thought. How could you call an enormous hole in the ground a "site"?—he gave a résumé of his earlier experiences there. Both listened attentively, and waited until he had finished before putting questions to him.

Marlow grew excited, more and more enthusiastic. His attention wandered from the winding, hilly roads that they now traversed. On numerous occasions he scraped the young trees and second-growth timber that pressed in on the lanes. But Daël betrayed no emotion, except for a speculative gleam in his eyes, as if the explanation of the mystery, the pure heart of it, the understanding and the knowledge, were all that mattered.

Conway liked him better because he didn't fathom the mathematician. He couldn't figure the man in black and white, but there was a definite comfort in Daël's presence on the trip, an aura of dispassionate, impersonal analysis.

And in the sky beyond, undimmed by the light of the moon that now stood well above the horizon, rose the wormlike pillar of flame, looming larger as they drew nearer, and, it seemed to Conway, unquestionably more monumental than it had been three hours ago.

When the car paused at the edge of the valley, he saw the rim of the crater dangerously close to the crest of the surrounding hills. The whole sphere must have increased its diameter by a quarter or a half mile since he had last seen it.

Above their heads, thrust high into the atmosphere, commenced the base of pale fire shaped like an inverted cone and capped with a voluted, splaying blur. He stared in fascination from the great flame vortex to the stupendous pit that was now more than a mile and a quarter in width and a half mile in depth. Projection in space of the full, true sphere would place the bottom of the flame at the exact top point of the sphere. And the edges of the earth rock dissolved before his gaze, soundlessly vanishing.

Professor Daël walked forward with a look of contemplative abstraction.

Marlow bustled among the instruments he had brought along and hastily set them up, Conway assisting him. The moonlighted darkness had begun to change in a ghostly, indefinable manner before Marlow was ready. Conway limped aside while the physicist took readings and measurements.

The twilight of dawn grew abroad, but as yet cast only a faintness of light, matching the pallid flame, when Marlow completed his field work. The physicist gave an exclamation of annoyance. There was no change in Daël's posture. He stood near the limits of the crater, retreating as it advanced.

"What's the trouble? Couldn't you find anything?" asked Conway. His voice reverberated curiously loud in the dead air.

Marlow snapped irritably, "The findings are either insufficient or impossible! According to the thermocouple, the flame has exactly the same temperature as the surrounding air—71.4° F; ash content, zero; no trace of gases; no emission of radiant energy. The walls of the crater in the field of destructive force are precisely at top soil temperature—62° F. It's utter nonsense. It makes a hopeless mystery!

"The encroachment of the pit would seem to be accelerating. I could calculate the exact figures after considerably more study, but roughly, the expansion started from nothing, from absolute inertia, and achieved a diameter of a mile in four days. It will be two miles at the end of the fifth day, six miles on the sixth, eighteen on the seventh, fifty-four on the eighth day, one hundred and sixty-two miles on the ninth day, and so on."

Conway whistled. "We've got to stop it!"

"Halt the total elimination or exhaustion of matter? How? Matter—rock, mineral, vegetation, everything, even the air—is disappearing into that sphere of space. It isn't being converted into heat, radiation, or any known form of energy. Some unknown force is operating. The flame offers evidence of the transformation, but the flame is mere candlelight compared to what it should be from the release of all the energy locked up in those millions of tons of matter that have vanished. The phenomenon violates the basic laws of physics."

Marlow walked toward the margin of the crater for a closer scrutiny.

Daël spoke sharply, "Marlow! Don't! Your life—"

But the physicist had already knelt by the dividing line, his head extended into the sphere of space.

Instantly the top of his skull disappeared as if sliced off. Surprise and horror briefly altered the remaining half of his face. His throat formed a full scream, but only a short, hoarse gasp issued from his lips as his body sagged forward.

The now headless torso toppled faster, a wave of invisibility seeming to sweep over it and to encompass it. Marlow's end was grotesque in the brightening dawn. His heels rose a foot from the ground. His shoes hung for a moment, soles up, the last earthly trace of him, before they followed his body into the void.

Conway's imagination played tricks on him. He thought that the flame quivered. He thought that he heard Marlow's voice sounding and resounding everywhere. Then the shock passed, and there was only the old, uncanny quiet.

Professor Daël turned away from the crater, a sadness on his face. He said, "I had just formed a hypothesis to explain what is happening here when Marlow's zeal carried him away. It's really rather simple. If only I could have warned him sooner—"

"Simple?"

"Why, yes. Our habits and modes of thought sometimes interfere with our ability to reason correctly. Otherwise I should have reached a conclusion in time to prevent my friend's death.

"We are accustomed to think of matter, whether it be a chair or a star, as something that exists in space. We regard any material object as possessing a definite and measurable length, breadth, and thickness. It displaces an identical volume of what would otherwise be empty space. Is that clear?"

"Yeah, so far. I'm surprised."

"Well, what is to prevent space from displacing matter?"

"Huh?" Conway was startled. He frowned while he tried to grasp the concept. "What the dickens do you mean? If space displaced matter, there wouldn't be any matter, would there? And where would it go?"

"For a mind that was not scientifically trained, yours has gone straight to the main points involved. However, it will be necessary to remind you of several theories concerning the nature and structure of space before I attempt to answer your questions.

"The former concept of an all-pervasive ether has largely been abandoned by modern scientists. They still consider it as a possibility in any thorough effort to grasp the nature of space, but have not discovered the slightest proof of its existence.

"On the other hand, no intellect that I have yet encountered has found it possible to form a visual image of space under the three principal alternative concepts, nor am I an exception. These alternatives are: that space is finite, or affected by a curvature that makes it return upon itself; that space is infinite; that space is finite but expanding. The human mind is unable to establish a mental picture of either of these three states.

"The usual objections are, to the first: if space is finite, what lies beyond the limits? To the second: how can space be infinite, and continue forever in all directions? To the third: why should space be expanding, and what would exist beyond the fringes of expansion?

"None of these concepts can be supported by tangible proof, hence the course of our reasoning must be theoretical."

"It's getting pretty deep for me, but I'm floundering along," said Conway wryly.

"Some mathematicians and physicists have considered time to be the fourth dimension of space, and from that theory arose much of the nonsense that used to be written about the possibility of travelling back and forth in time just as we move about in space. The scientists had a definite, specialized meaning for 'dimension' as a word concept. The popular mind associated dimension with the length, breadth, and thickness of objects, and thus confused time-dimension with space-dimension.

"To correct the popular error, time is not a dimension at all. It is a quality attending the existence of space and matter. It is a unit by which we measure the continued existence or duration of space and matter, of life and inanimate objects. If the public mind had been able to reason clearly, to use accurate definitions, and to emphasize distinctions, it would never have subscribed to the fallacy of time travelling."

After hard thinking, Conway asked, "Then, if we stick to the popular use of words, it would straighten itself out as a three-dimensional universe? And time would be off in another category, as a sort of necessary quality or attribute? That is, the three-dimensional universe couldn't exist without time, but time is a different yardstick by itself; whereas length, breadth, and thickness are three dimensions that belong together?"

"Yes."

"O.K. So we have a universe with three space dimensions and one time unit. Go on."

"Now, the great difficulty that has stood in the way of past attempts to understand the universe is that men have tried to visualize the whole by means of its parts. Most persons think of space and the universe as a sort of gigantic ball or sphere or balloon. They form a hazy opinion of the whole by thinking in the habitual terms of the world to which they are accustomed.

"Under the old theory of a time-space continuum, this universe was approached from a four-dimensional viewpoint; but as I have just pointed out, the fourth dimension was time, which we have now transferred to a separate category.

"Next, let us leap out of our mental ruts. Whether space is finite, or infinite, or expanding is not the essential truth and knowledge that we seek. The essential truth is that, whichever of these concepts happens to be valid, there must necessarily exist a condition that we cannot understand because we are not there to see it and cannot visualize because it is greater than the universe of three space dimensions and one time unit.

"This state or condition must therefore be a true fourth dimension in addition to the other three dimensions and the time factor.

"That new, fourth dimension exists only beyond space, if space is finite or expanding, and exists only beyond infinity if space is infinite. It does not exist anywhere else, and is not present anywhere within the universe of three space dimensions and one time unit. It cannot exist within the known cosmos."

"What's this fourth dimension like? How does it affect the universe?"

"That I can answer only in part. You are looking at the effects of that dimension, however."

"What! You mean the crater? But you just said the dimension couldn't exist here!"

Daël nodded. "Think a moment and you will see what I mean. That dimension does not and cannot exist here. Therefore, it is eliminating matter, with its three space dimensions and single time factor. We are viewing the effects of that process."

"But what started it? Why should it happen here?" protested Conway.

"It could be called an inevitable development, but I believe that it came as a direct result of the bomb that destroyed the U.C.T.L.

That building contained all the elements in the universe, all the compounds organic and inorganic known to science. At the moment of explosion, the whole universe was concentrated here in miniature, thus creating that new, fourth dimension of ultra-space.

"And since ultra-space cannot exist in our universe, the ultra-space from beyond instantly curved in to unite with the segment of ultra-space here. Together they are erasing, blotting out, eliminating the three space dimensions and the time factor. Our universe is vanishing. Space, matter, energy, time, and life are ceasing."

Conway, stunned, managed to gasp, "Why wasn't the whole business over in a jiffy, then?"

"It was!" came the electrifying answer. "From the viewpoint of our time factor, the process will be cumulative. It will be weeks before Earth vanishes, months and years before the entire universe vanishes. But if we could instantaneously transport ourselves into ultra-space, we would find that nothing existed except ultra-space with its fourth dimension and possibly additional dimensions that are wholly beyond my imagination. We would find no trace of this universe, and we would discover that we ourselves had died long ago."

A nervous tremor shook Conway. It seemed to him that the margin of the expanding sphere was creeping and flowing, faster and faster, at ever-increasing speed. And toward the sky, lightening with dawn, rose the vast, pallid, inert vortex of flame, like a finger of destiny, the sword of doom.

He couldn't linger. He didn't want to linger. He had to get back to his desk, send out bulletins.

Newark, Aug. 8 (WP)—The great crater discovered and described yesterday by Walter Conway is now two miles wide and still growing, according to the same eyewitnesses. The crater is locally known as the "flame terror" because of a strange fire above it. The expansion of the crater follows the predictions of the physicist, C. H. Marlow, who was killed while investigating it.

Newark, Aug. 9 (WP)—The great crater in northern New Jersey is now six miles in diameter. No means have been found to check its onrush.

To his previously published theory about its origin, Professor Daël today added that the crater is proof that a straight line would not meet itself at infinity, as mathematicians have hitherto believed, but that it would be absorbed and hence would disappear in ultra-space.

Newark, Aug. 10 (WP)—The great crater is now more than eighteen miles in diameter. Evacuation of near-by towns has begun.

U. S. Govt. HQ., Aug 10 (WP)—It was officially announced that an armistice had been declared by all warring powers effective at noon today (E.S.T.).

The proclamation comes as a complete surprise. Political observers are unable to account for the swiftness and unanimity of action by the hostile nations.

New York, Aug. 11 (WP)—Scientists from all over the world are speeding here by stratoplane to observe the great crater which today reached a diameter of fifty-four miles. New York is being evacuated amid scenes of wildest disorder. Among the victims was Professor Daël, famous mathematician, who was killed by an infuriated mob when he stated that the sphere of ultra-space would expand to embrace the globe within two weeks, and thus eliminate every trace of man's existence.

—1941—

THE LIBRARY OF BABEL

Jorge Luis Borges

Born in Argentina, which formed the setting and inspiration for much of his fiction, Jorge Luis Borges (1899–1986) is acclaimed the world over for his unique imagination and philosophically reflective prose. His work consistently explores themes of time, infinity and randomness, and a number of books on his use and knowledge of mathematics have appeared over the years. The only author in this collection to have been shortlisted for the Nobel Prize for literature, Borges may seem a far cry from the likes of Lovecraft, Long, and Wandrei, but he was certainly an admirer of the spirit of the weird. A late story, "There are More Things" (1975), is dedicated "To the memory of H. P. Lovecraft", and tells of the amorphous new occupier of a house in South America. An enjoyable pastiche, the story is also notable for its direct references to Charles Hinton's four-dimensional geometry. The story selected here, however, is perhaps his most famous and beloved, and asks the reader to ponder the nature of infinity in a geometrically symmetrical world-library. It first appeared in his 1941 collection, El jardín de senderos que se bifurcan *(*The Garden of Forking Paths*), with English language translations arriving twenty-one years later in 1962.*

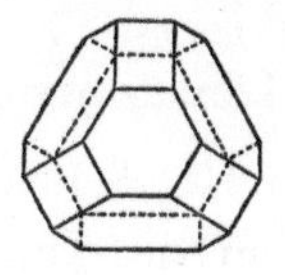

By this art you may contemplate the variation of the 23 letters...

ANATOMY OF MELANCHOLY, PT. 2, SEC. II, MEM. IV

THE UNIVERSE (WHICH OTHERS CALL THE LIBRARY) IS COMposed of an indefinite, perhaps infinite number of hexagonal galleries. In the centre of each gallery is a ventilation shaft, bounded by a low railing. From any hexagon one can see the floors above and below—one after another, endlessly. The arrangement of the galleries is always the same: Twenty bookshelves, five to each side, line four of the hexagon's six sides; the height of the bookshelves, floor to ceiling, is hardly greater than the height of a normal librarian. One of the hexagon's free sides opens onto a narrow sort of vestibule, which in turn opens onto another gallery, identical to the first—identical in fact to all. To the left and right of the vestibule are two tiny compartments. One is for sleeping, upright; the other, for satisfying one's physical necessities. Through this space, too, there passes a spiral staircase, which winds upward and downward into the remotest distance. In the vestibule there is a mirror, which faithfully duplicates appearances. Men often infer from this mirror that the Library is not infinite—if it were, what need would there be for that illusory replication? I prefer to dream that burnished surfaces are a figuration and promise of the infinite... Light is provided by certain spherical fruits that bear the name "bulbs." There are two of these bulbs in each hexagon, set crosswise. The light they give is insufficient, and unceasing.

Like all the men of the Library, in my younger days I travelled; I have journeyed in quest of a book, perhaps the catalogue of catalogues. Now that my eyes can hardly make out what I myself have written, I am preparing to die, a few leagues from the hexagon where I was born. When I am dead, compassionate hands will throw me over the railing; my tomb will be the unfathomable air, my body will sink for ages, and will decay and dissolve in the wind engendered by my fall, which shall be infinite. I declare that the Library is endless. Idealists argue that the hexagonal rooms are the necessary shape of absolute space, or at least of our *perception* of space. They argue that a triangular or pentagonal chamber is inconceivable. (Mystics claim that their ecstasies reveal to them a circular chamber containing an enormous circular book with a continuous spine that goes completely around the walls. But their testimony is suspect, their words obscure. That cyclical book is God.) Let it suffice for the moment that I repeat the classic dictum: *The Library is a sphere whose exact centre is any hexagon and whose circumference is unattainable.*

Each wall of each hexagon is furnished with five bookshelves; each bookshelf holds thirty-two books identical in format; each book contains four hundred ten pages; each page, forty lines; each line, approximately eighty black letters. There are also letters on the front cover of each book; those letters neither indicate nor prefigure what the pages inside will say. I am aware that that lack of correspondence once struck men as mysterious. Before summarizing the solution of the mystery (whose discovery, in spite of its tragic consequences, is perhaps the most important event in all history), I wish to recall a few axioms.

First: *The Library has existed* ab æternitate. That truth, whose immediate corollary is the future eternity of the world, no rational mind can doubt. Man, the imperfect librarian, may be the work of chance or of malevolent demiurges; the universe, with its elegant appointments—its bookshelves, its enigmatic books, its indefatigable staircases for the traveller, and its water closets for the seated librarian—can only be the handiwork of a god. In order to grasp the distance that separates the human and the divine, one has only to compare these crude trembling symbols which my fallible hand scrawls on the cover of a book with the organic letters inside—neat, delicate, deep black, and inimitably symmetrical.

Second: *There are twenty-five orthographic symbols.** That discovery enabled mankind, three hundred years ago, to formulate a general theory of the Library and thereby satisfactorily solve the riddle that no conjecture had been able to divine—the formless and chaotic nature of virtually all books. One book, which my father once saw in a hexagon in circuit 15-94, consisted of the letters M C V perversely repeated from the first line to the last. Another (much consulted in this zone) is a mere labyrinth of letters whose penultimate page contains the phrase *O Time thy pyramids*. This much is known: For every rational line or forthright statement there are leagues of senseless cacophony, verbal nonsense, and incoherency. (I know of one semibarbarous zone whose librarians repudiate the "vain and superstitious habit" of trying to find sense in books, equating such a quest with attempting to find meaning in dreams or in the chaotic lines of the palm of one's hand… They will acknowledge that the

* The original manuscript has neither numbers nor capital letters; punctuation is limited to the comma and the period. Those two marks, the space, and the twenty-two letters of the alphabet are the twenty-five sufficient symbols that our unknown author is referring to. [Ed. note.]

inventors of writing imitated the twenty-five natural symbols, but contend that that adoption was fortuitous, coincidental, and that books in themselves have no meaning. That argument, as we shall see, is not entirely fallacious.)

For many years it was believed that those impenetrable books were in ancient or far-distant languages. It is true that the most ancient peoples, the first librarians, employed a language quite different from the one we speak today; it is true that a few miles to the right, our language devolves into dialect and that ninety floors above, it becomes incomprehensible. All of that, I repeat, is true—but four hundred ten pages of unvarying M C V's cannot belong to any language, however dialectal or primitive it may be. Some have suggested that each letter influences the next, and that the value of M C V on page 71, line 3, is not the value of the same series on another line of another page, but that vague thesis has not met with any great acceptance. Others have mentioned the possibility of codes; that conjecture has been universally accepted, though not in the sense in which its originators formulated it.

Some five hundred years ago, the chief of one of the upper hexagons* came across a book as jumbled as all the others, but containing almost two pages of homogeneous lines. He showed his find to a travelling decipherer, who told him that the lines were written in Portuguese; others said it was Yiddish. Within the

* In earlier times, there was one man for every three hexagons. Suicide and diseases of the lung have played havoc with that proportion. An unspeakably melancholy memory: I have sometimes travelled for nights on end, down corridors and polished staircases, without coming across a single librarian.

century experts had determined what the language actually was: a Samoyed-Lithuanian dialect of Guaraní, with inflections from classical Arabic. The content was also determined: the rudiments of combinatory analysis, illustrated with examples of endlessly repeating variations. Those examples allowed a librarian of genius to discover the fundamental law of the Library. This philosopher observed that all books, however different from one another they might be, consist of identical elements: the space, the period, the comma, and the twenty-two letters of the alphabet. He also posited a fact which all travellers have since confirmed: *In all the Library, there are no two identical books*. From those incontrovertible premises, the librarian deduced that the Library is "total"—perfect, complete, and whole—and that its bookshelves contain all possible combinations of the twenty-two orthographic symbols (a number which, though unimaginably vast, is not infinite)—that is, all that is able to be expressed, in every language. *All*—the detailed history of the future, the autobiographies of the archangels, the faithful catalogue of the Library, thousands and thousands of false catalogues, the proof of the falsity of those false catalogues, a proof of the falsity of the *true* catalogue, the gnostic gospel of Basilides, the commentary upon that gospel, the commentary on the commentary on that gospel, the true story of your death, the translation of every book into every language, the interpolations of every book into all books, the treatise Bede could have written (but did not) on the mythology of the Saxon people, the lost books of Tacitus.

When it was announced that the Library contained all books, the first reaction was unbounded joy. All men felt themselves the possessors of an intact and secret treasure. There was no personal problem, no world problem, whose eloquent solution did not exist—somewhere in some hexagon. The universe was justified; the universe suddenly became congruent with the unlimited width and breadth of humankind's hope. At that period there was much talk of The Vindications—books of *apologiæ* and prophecies that would vindicate for all time the actions of every person in the universe and that held wondrous arcana for men's futures. Thousands of greedy individuals abandoned their sweet native hexagons and rushed downstairs, upstairs, spurred by the vain desire to find their Vindication. These pilgrims squabbled in the narrow corridors, muttered dark imprecations, strangled one another on the divine staircases, threw deceiving volumes down ventilation shafts, were themselves hurled to their deaths by men of distant regions. Others went insane... The Vindications do exist (I have seen two of them, which refer to persons in the future, persons perhaps not imaginary), but those who went in quest of them failed to recall that the chance of a man's finding his own Vindication, or some perfidious version of his own, can be calculated to be zero.

At that same period there was also hope that the fundamental mysteries of mankind—the origin of the Library and of time—might be revealed. In all likelihood those profound mysteries can indeed be explained in words; if the language of the philosophers is not sufficient, then the multiform Library must surely have produced the extraordinary language that is required, together with the words and grammar of that language. For four centuries, men have been scouring the hexagons... There are official searchers, the "inquisitors." I have seen them about their tasks: they arrive exhausted at some hexagon, they talk about a staircase that nearly killed them—rungs were missing—they speak with the librarian about galleries and staircases, and, once in a while, they take up the nearest book and leaf through it, searching for disgraceful or dishonourable words. Clearly, no one expects to discover anything.

That unbridled hopefulness was succeeded, naturally enough, by a similarly disproportionate depression. The certainty that some bookshelf in some hexagon contained precious books, yet that those precious books were forever out of reach, was almost unbearable. One blasphemous sect proposed that the searches be discontinued and that all men shuffle letters and symbols until those canonical books, through some improbable stroke of chance, had been constructed. The authorities were forced to issue strict orders. The sect disappeared, but in my childhood I have seen old men who for long periods would hide in the latrines with metal disks and a forbidden dice cup, feebly mimicking the divine disorder.

Others, going about it in the opposite way, thought the first thing to do was eliminate all worthless books. They would invade the hexagons, show credentials that were not always false, leaf disgustedly through a volume, and condemn entire walls of books. It is to their hygienic, ascetic rage that we lay the senseless loss of millions of volumes. Their name is execrated today, but those who grieve over the "treasures" destroyed in that frenzy overlook two widely acknowledged facts: One, that the Library is so huge that any reduction by human hands must be infinitesimal. And two, that each book is unique and irreplaceable, but (since the Library is total) there are always several hundred thousand imperfect facsimiles—books that differ by no more than a single letter, or a comma. Despite general opinion, I daresay that the consequences of the depredations committed by the Purifiers have been exaggerated by the horror those same fanatics inspired. They were spurred on by the holy zeal to reach—someday, through unrelenting effort—the books of the Crimson Hexagon—books smaller than natural books, books omnipotent, illustrated, and magical.

We also have knowledge of another superstition from that period: belief in what was termed the Book-Man. On some shelf in some hexagon, it was argued, there must exist a book that is the cipher and perfect compendium *of all other books*, and some librarian must have examined that book; this librarian is analogous to a god. In the language of this zone there are still vestiges of the sect that worshipped that distant librarian. Many have gone in search of Him. For a hundred years, men beat every possible path—and every path in vain. How was one to locate the idolized secret hexagon that sheltered Him? Someone proposed searching by regression: To locate book A, first consult book B, which tells where book A can be found; to locate book B, first consult book C, and so on, to infinity... It is in ventures such as these that I have squandered and spent my years. I cannot think it unlikely that there is such a total book* on some shelf in the universe. I pray to the unknown gods that some man—even a single man, tens of centuries ago—has perused and read that book. If the honour and wisdom and joy of such a reading are not to be my own, then let them be for others. Let heaven exist, though my own place be in hell. Let

* I repeat: In order for a book to exist, it is sufficient that it be *possible*. Only the impossible is excluded. For example, no book is also a staircase, though there are no doubt books that discuss and deny and prove that possibility, and others whose structure corresponds to that of a staircase.

me be tortured and battered and annihilated, but let there be one instant, one creature, wherein thy enormous Library may find its justification.

Infidels claim that the rule in the Library is not "sense," but "non-sense," and that "rationality" (even humble, pure coherence) is an almost miraculous exception. They speak, I know, of "the feverish Library, whose random volumes constantly threaten to transmogrify into others, so that they affirm all things, deny all things, and confound and confuse all things, like some mad and hallucinating deity." Those words, which not only proclaim disorder but exemplify it as well, prove, as all can see, the infidels' deplorable taste and desperate ignorance. For while the Library contains all verbal structures, all the variations allowed by the twenty-five orthographic symbols, it includes not a single absolute piece of nonsense. It would be pointless to observe that the finest volume of all the many hexagons that I myself administer is titled *Combed Thunder*, while another is titled *The Plaster Cramp*, and another, *Axaxaxas mlö*. Those phrases, at first apparently incoherent, are undoubtedly susceptible to cryptographic or allegorical "reading"; that reading, that justification of the words' order and existence, is itself verbal and, *ex hypothesi*, already contained somewhere in the Library. There is no combination of characters one can make—*dhcmrlchtdj*, for example—that the divine Library has not foreseen and that in one or more of its secret tongues does not hide a terrible significance. There is no syllable one can speak that is not filled with tenderness and terror, that is not, in one of those languages, the mighty name of a god. To speak is to commit tautologies. This pointless, verbose epistle already exists in one of the thirty volumes of the five bookshelves in one of the countless hexagons—as does its refutation. (A number *n* of the possible languages employ the same vocabulary; in some of them, the *symbol* "library" possesses the correct definition "everlasting, ubiquitous system of hexagonal galleries," while a library—the thing—is a loaf of bread or a pyramid or something else, and the six words that define it themselves have other definitions. You who read me—are you certain you understand my language?)

Methodical composition distracts me from the present condition of humanity. The certainty that everything has already been written annuls us, or renders us phantasmal. I know districts in which the young people prostrate themselves before books and like savages kiss their pages, though they cannot read a letter. Epidemics, heretical discords, pilgrimages that inevitably degenerate into brigandage have decimated the population. I believe I mentioned the suicides, which are more and more frequent every year. I am perhaps misled by old age and fear, but I suspect that the human species—the *only* species—teeters at the verge of extinction, yet that the Library—enlightened, solitary, infinite, perfectly unmoving, armed with precious volumes, pointless, incorruptible, and secret—will endure.

I have just written the word "infinite." I have not included that adjective out of mere rhetorical habit; I hereby state that it is not illogical to think that the world is infinite. Those who believe it to have limits hypothesize that in some remote place or places the corridors and staircases and hexagons may, inconceivably, end—which is absurd. And yet those who picture the world as unlimited forget that the number of possible books is *not*. I will be bold enough to suggest this solution to the ancient problem: *The Library is unlimited but periodic*. If an eternal traveller should journey in any direction, he would find after untold centuries that the same volumes are repeated in the same disorder—which, repeated, becomes order: the Order. My solitude is cheered by that elegant hope.*

Mar del Plata, 1941

* Letizia Alvarez de Toledo has observed that the vast Library is pointless; strictly speaking, all that is required is *a single volume*, of the common size, printed in nine- or ten-point type, that would consist of an infinite number of infinitely thin pages. (In the early seventeenth century, Cavalieri stated that every solid body is the superposition of an infinite number of planes.) Using that silken *vademecum* would not be easy: each apparent page would open into other similar pages; the inconceivable middle page would have no "back."

—1941—

"—AND HE BUILT A CROOKED HOUSE—"

Robert Heinlein

One of the so-called "Big Three" of science fiction—along with his contemporaries Isaac Asimov and Arthur C. Clarke—Robert Anson Heinlein (1907–1988) is perhaps best remembered for two novels, Starship Troopers *(1959) and* Stranger in a Strange Land *(1961). Abandoning a brief career in the navy after a case of pulmonary tuberculosis left him unfit to serve, Heinlein turned his hand to aviation engineering before the success of his later novels allowed him to take up writing full time. This story is from his early period and was first published in* Astounding Science Fiction, *February 1941. An architectural exploration of Hinton's tessaracts, the story poses a seemingly simple question: "What is a house?"*

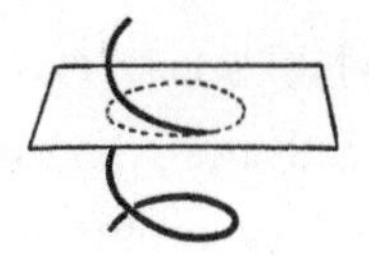

AMERICANS ARE CONSIDERED CRAZY ANYWHERE IN THE world.

They will usually concede a basis for the accusation but point to California as the focus of the infection. Californians stoutly maintain that their bad reputation is derived solely from the acts of the inhabitants of Los Angeles County. Angelenos will, when pressed, admit the charge but explain hastily, "It's Hollywood. It's not our fault—we didn't ask for it; Hollywood just grew."

The people in Hollywood don't care; they glory in it. If you are interested, they will drive you up Laurel Canyon "—where we keep the violent cases." The Canyonites—the brown-legged women, the trunks-clad men constantly busy building and rebuilding their slap-happy unfinished houses—regard with faint contempt the dull creatures who live down in the flats, and treasure in their hearts the secret knowledge that they, and only they, know how to live.

Lookout Mountain Avenue is the name of a side canyon which twists up from Laurel Canyon. The other Canyonites don't like to have it mentioned; after all, one must draw the line somewhere!

High up on Lookout Mountain at number 8775, across the street from the Hermit—the original Hermit of Hollywood—lived Quintus Teal, graduate architect.

Even the architecture of southern California is different. Hot dogs are sold from a structure built like and designated "The Pup." Ice cream cones come from a giant stucco ice cream cone, and neon proclaims "Get the Chili Bowl Habit!" from the roofs of buildings which are indisputably chilli bowls. Gasoline, oil, and free road maps are dispensed beneath the wings of tri-motored transport planes, while the certified rest rooms, inspected hourly for your comfort, are located in the cabin of the plane itself. These things may surprise, or amuse, the tourist, but the local residents, who walk bare-headed in the famous California noonday sun, take them as a matter of course.

Quintus Teal regarded the efforts of his colleagues in architecture as faint-hearted, fumbling, and timid.

"What is a house?" Teal demanded of his friend, Homer Bailey.

"Well—" Bailey admitted cautiously, "speaking in broad terms, I've always regarded a house as a gadget to keep off the rain."

"Nuts! You're as bad as the rest of them."

"I didn't say the definition was complete—"

"Complete! It isn't even in the right direction. From that point of view we might just as well be squatting in caves. But I don't blame you," Teal went on magnanimously, "you're no worse than the lugs you find practising architecture. Even the Moderns—all they've done is to abandon the Wedding Cake School in favour of the Service Station School, chucked away the gingerbread and slapped on some chromium, but at heart they are as conservative and traditional as a county courthouse. Neutra! Schindler! What have those bums got? What's Frank Lloyd Wright got that I haven't got?"

"Commissions," his friend answered succinctly.

"Huh? Wha' d'ju say?" Teal stumbled slightly in his flow of words, did a slight double take, and recovered himself. "Commissions. Correct. And why? Because I don't think of a house as an upholstered cave; I think of it as a machine for living, a vital process, a live dynamic thing, changing with the mood of the dweller—not a dead, static, oversized coffin. Why should we be held down by the frozen concepts of our ancestors? Any fool with a little smattering of descriptive geometry can design a house in the ordinary way. Is the static geometry of Euclid the only mathematics? Are we to completely disregard the Picard-Vessiot theory? How about modular systems?—to say nothing of the rich suggestions of stereochemistry. Isn't there a place in architecture for transformation, for homomorphology, for actional structures?"

"Blessed if I know," answered Bailey. "You might just as well be talking about the fourth dimension for all it means to me."

"And why not? Why should we limit ourselves to the— Say!" He interrupted himself and stared into distances. "Homer, I think you've really got something. After all, why not? Think of the infinite richness of articulation and relationship in four dimensions. What a house, what a house—" He stood quite still, his pale bulging eyes blinking thoughtfully.

Bailey reached up and shook his arm. "Snap out of it. What the hell are you talking about, four dimensions? Time is the fourth dimension; you can't drive nails into *that*."

Teal shrugged him off. "Sure. Sure. Time is *a* fourth dimension, but I'm thinking about a fourth spatial dimension, like length, breadth and thickness. For economy of materials and convenience of arrangement you couldn't beat it. To say nothing of the saving of ground space—you could put an eight-room house on the land now occupied by a one-room house. Like a tesseract—"

"What's a tesseract?"

"Didn't you go to school? A tesseract is a hypercube, a square figure with four dimensions to it, like a cube has three, and a square has two. Here, I'll show you." Teal dashed out into the kitchen of his apartment and returned with a box of toothpicks which he spilled on the table between them, brushing glasses and a nearly empty Holland gin bottle carelessly aside. "I'll need some plasticine. I had some around here last week." He burrowed into a drawer of the littered desk which crowded one corner of his dining room and emerged with a lump of oily sculptor's clay. "Here's some."

"What are you going to do?"

"I'll show you." Teal rapidly pinched off small masses of the clay and rolled them into pea-sized balls. He stuck toothpicks into four of these and hooked them together into a square. "There! That's a square."

"Obviously."

"Another one like it, four more toothpicks, and we make a cube." The toothpicks were now arranged in the framework of a square box, a cube, with the pellets of clay holding the corners together. "Now we make another cube just like the first one, and the two of them will be two sides of the tesseract."

Bailey started to help him roll the little balls of clay for the second cube, but became diverted by the sensuous feel of the docile clay and started working and shaping it with his fingers.

"Look," he said, holding up his effort, a tiny figurine, "Gypsy Rose Lee."

"Looks more like Gargantua; she ought to sue you. Now pay attention. You open up one corner of the first cube, interlock the second cube at one corner, and then close the corner. Then take eight more toothpicks and join the bottom of the first cube to the bottom of the second, on a slant, and the top of the first to the top of the second, the same way." This he did rapidly, while he talked.

"What's that supposed to be?" Bailey demanded suspiciously.

"That's a tesseract, eight cubes forming the sides of a hypercube in four dimensions."

"It looks more like a cat's cradle to me. You've only got two cubes there anyhow. Where are the other six?"

"Use your imagination, man. Consider the top of the first cube in relation to the top of the second; that's cube number three. Then the two bottom squares, then the front faces of each cube, the back faces, the right hand, the left hand—eight cubes." He pointed them out.

"Yeah, I see 'em. But they still aren't cubes; they're whatchamu-callems—prisms. They are not square, they slant."

"That's just the way you look at it, in perspective. If you drew a picture of a cube on a piece of paper, the side squares would be slaunchwise, wouldn't they? That's perspective. When you look at a four-dimensional figure in three dimensions, naturally it looks crooked. But those are all cubes just the same."

"Maybe they are to you, brother, but they still look crooked to me."

Teal ignored the objections and went on. "Now consider this as the framework of an eight-room house; there's one room on the ground floor—that's for service, utilities, and garage. There are six rooms opening off it on the next floor, living room, dining room, bath, bedrooms, and so forth. And up at the top, completely inclosed and with windows on four sides, is your study. There! How do you like it?"

"Seems to me you have the bathtub hanging out of the living room ceiling. Those rooms are interlaced like an octopus."

"Only in perspective, only in perspective. Here, I'll do it another way so you can see it." This time Teal made a cube of toothpicks, then made a second of halves of toothpicks, and set it exactly in the centre of the first by attaching the corners of the small cube to the large cube by short lengths of toothpick. "Now—the big cube is your ground floor, the little cube inside is your study on the top floor. The six cubes joining them are the living rooms. See?"

Bailey studied the figure, then shook his head. "I still don't see but two cubes, a big one and a little one. Those other six things, they look like pyramids this time instead of prisms, but they still aren't cubes."

"Certainly, certainly, you are seeing them in different perspective. Can't you see that?"

"Well, maybe. But that room on the inside, there. It's completely surrounded by the thingamujigs. I thought you said it had windows on four sides."

"It has—it just looks like it was surrounded. That's the grand feature about a tesseract house, complete outside exposure for every room, yet every wall serves two rooms and an eight-room house requires only a one-room foundation. It's revolutionary."

"That's putting it mildly. You're crazy, bud; you can't build a house like that. That inside room is on the inside, and there she stays."

Teal looked at his friend in controlled exasperation. "It's guys like you that keep architecture in its infancy. How many square sides has a cube?"

"Six."

"How many of them are inside?"

"Why, none of 'em. They're all on the outside."

"All right. Now listen—a tesseract has eight cubical sides, *all on the outside*. Now watch me. I'm going to open up this tesseract like you can open up a cubical pasteboard box, until it's flat. That way you'll be able to see all eight of the cubes." Working very rapidly he constructed four cubes, piling one on top of the other in an unsteady tower. He then built out four more cubes from the four exposed faces of the second cube in the pile. The structure swayed a little under the loose coupling of the clay pellets, but it stood, eight cubes in an inverted cross, a double cross, as the four additional cubes stuck out in four directions. "Do you see it now? It rests on the ground floor room, the next six cubes are the living rooms, and there is your study, up at the top."

Bailey regarded it with more approval than he had the other figures. "At least I can understand it. You say that is a tesseract, too?"

"That is a tesseract unfolded in three dimensions. To put it back together you tuck the top cube onto the bottom cube, fold those side cubes in till they meet the top cube and there you are. You do all this folding through a fourth dimension of course; you don't distort any of the cubes, or fold them into each other."

Bailey studied the wobbly framework further. "Look here," he said at last, "why don't you forget about folding this thing up through a fourth dimension—you can't anyway—and build a house like this?"

"What do you mean, I can't? It's a simple mathematical problem—"

"Take it easy, son. It may be simple in mathematics, but you could never get your plans approved for construction. There isn't any fourth dimension; forget it. But this kind of a house—it might have some advantages."

Checked, Teal studied the model. "Hm-m-m— Maybe you got something. We could have the same number of rooms, and we'd save the same amount of ground space. Yes, and we would set that middle cross-shaped floor northeast, southwest, and so forth, so that every room would get sunlight all day long. That central axis lends itself nicely to central heating. We'll put the dining room on the northeast and the kitchen on the southeast, with big view windows in every room. O.K., Homer, I'll do it! Where do you want it built?"

"Wait a minute! Wait a minute! I didn't say you were going to build it for me—"

"Of course I am. Who else? Your wife wants a new house; this is it."

"But Mrs. Bailey wants a Georgian house—"

"Just an idea she has. Women don't know what they want—"

"Mrs. Bailey does."

"Just some idea an out-of-date architect has put in her head. She drives a 1941 car, doesn't she? She wears the very latest styles—why should she live in an eighteenth century house? This house will be even later than a 1941 model; it's years in the future. She'll be the talk of the town."

"Well—I'll have to talk to her."

"Nothing of the sort. We'll surprise her with it. Have another drink."

"Anyhow, we can't do anything about it now. Mrs. Bailey and I are driving up to Bakersfield tomorrow. The company's bringing in a couple of wells tomorrow."

"Nonsense. That's just the opportunity we want. It will be a surprise for her when you get back. You can just write me a cheque right now, and your worries are over."

"I oughtn't to do anything like this without consulting her. She won't like it."

"Say, who wears the pants in your family anyhow?"

The cheque was signed about halfway down the second bottle.

Things are done fast in southern California. Ordinary houses there are usually built in a month's time. Under Teal's impassioned heckling the tesseract house climbed dizzily skyward in days rather than weeks, and its cross-shaped second storey came jutting out at the four corners of the world. He had some trouble at first with the inspectors over these four projecting rooms but by using strong girders and folding money he had been able to convince them of the soundness of his engineering.

By arrangement, Teal drove up in front of the Bailey residence the morning after their return to town. He improvised on his two-tone horn. Bailey stuck his head out the front door. "Why don't you use the bell?"

"Too slow," answered Teal cheerfully. "I'm a man of action. Is Mrs. Bailey ready? Ah, there you are, Mrs. Bailey! Welcome home, welcome home. Jump in, we've got a surprise for you!"

"You know Teal, my dear," Bailey put in uncomfortably.

Mrs. Bailey sniffed. "I know him. We'll go in our own car, Homer."

"Certainly, my dear."

"Good idea," Teal agreed; " 'sgot more power than mine; we'll get there faster. I'll drive, I know the way." He took the keys from Bailey, slid into the driver's seat, and had the engine started before Mrs. Bailey could rally her forces.

"Never have to worry about my driving," he assured Mrs. Bailey, turning his head as he did so, while he shot the powerful car down the avenue and swung onto Sunset Boulevard, "it's a matter of power and control, a dynamic process, just my meat—I've never had a serious accident."

"You won't have but one," she said bitingly. "Will you *please* keep your eyes on the traffic?"

He attempted to explain to her that a traffic situation was a matter, not of eyesight, but intuitive integration of courses, speeds, and probabilities, but Bailey cut him short. "Where is the house, Quintus?"

"House?" asked Mrs. Bailey suspiciously. "What's this about a house, Homer? Have you been up to something without telling me?"

Teal cut in with his best diplomatic manner. "It certainly is a house, Mrs. Bailey. And what a house! It's a surprise for you from a devoted husband. Just wait till you see it—"

"I shall," she agreed grimly. "What style is it?"

"This house sets a new style. It's later than television, newer than next week. It must be seen to be appreciated. By the way," he went on rapidly, heading off any retort, "did you folks feel the earthquake last night?"

"Earthquake? What earthquake? Homer, was there an earthquake?"

"Just a little one," Teal continued, "about two a. m. If I hadn't been awake, I wouldn't have noticed it."

Mrs. Bailey shuddered. "Oh, this awful country! Do you hear that, Homer? We might have been killed in our beds and never have known it. Why did I ever let you persuade me to leave Iowa?"

"But my dear," he protested hopelessly, "you wanted to come out to California; you didn't like Des Moines."

"We needn't go into that," she said firmly. "You are a man; you should anticipate such things. Earthquakes!"

"That's one thing you needn't fear in your new home, Mrs. Bailey," Teal told her. "It's absolutely earthquake-proof; every part is in perfect dynamic balance with every other part."

"Well, I hope so. Where is this house?"

"Just around this bend. There's the sign now." A large arrow sign, of the sort favoured by real estate promoters, proclaimed in letters that were large and bright even for southern California:

THE HOUSE OF THE FUTURE!!!

COLOSSAL—AMAZING—REVOLUTIONARY

SEE HOW YOUR GRANDCHILDREN WILL LIVE!

Q. Teal, Architect

"Of course that will be taken down," he added hastily, noting her expression, "as soon as you take possession." He slued around the corner and brought the car to a squealing halt in front of the House of the Future. "*Voilà!*" He watched their faces for response.

Bailey stared unbelievingly, Mrs. Bailey in open dislike. They saw a simple cubical mass, possessing doors and windows, but no other architectural features, save that it was decorated in intricate mathematical designs. "Teal," Bailey asked slowly, "what have you been up to?"

Teal turned from their faces to the house. Gone was the crazy tower with its jutting second-storey rooms. No trace remained of the seven rooms above ground floor level. Nothing remained but the single room that rested on the foundations. "Great jumping cats!" he yelled, "I've been robbed!"

He broke into a run.

But it did him no good. Front or back, the story was the same: the other seven rooms had disappeared, vanished completely. Bailey caught up with him, and took his arm. "Explain yourself. What is this about being robbed? How come you built anything like this—it's not according to agreement."

"But I didn't. I built just what we had planned to build, an eight-room house in the form of a developed tesseract. I've been sabotaged; that's what it is! Jealousy! The other architects in town didn't dare let me finish this job; they knew they'd be washed up if I did."

"When were you last here?"

"Yesterday afternoon."

"Everything all right then?"

"Yes. The gardeners were just finishing up."

Bailey glanced around at the faultlessly manicured landscaping. "I don't see how seven rooms could have been dismantled and carted away from here in a single night without wrecking this garden."

Teal looked around, too. "It doesn't look it. I don't understand it."

Mrs. Bailey joined them. "Well? Well? Am I to be left to amuse myself? We might as well look it over as long as we are here, though I'm warning you, Homer, I'm not going to like it."

"We might as well," agreed Teal, and drew a key from his pocket with which he let them in the front door. "We may pick up some clues."

The entrance hall was in perfect order, the sliding screens that separated it from the garage space were back, permitting them to see the entire compartment. "This looks all right," observed Bailey. "Let's go up on the roof and try to figure out what happened. Where's the staircase? Have they stolen that, too?"

"Oh, no," Teal denied, "look—" He pressed a button below the light switch; a panel in the ceiling fell away and a light, graceful flight of stairs swung noiselessly down. Its strength members were the frosty silver of duralumin, its treads and risers transparent plastic. Teal wriggled like a boy who has successfully performed a card trick, while Mrs. Bailey thawed perceptibly.

It was beautiful.

"Pretty slick," Bailey admitted. "Howsomever it doesn't seem to go any place—"

"Oh, that—" Teal followed his gaze. "The cover lifts up as you approach the top. Open stair wells are anachronisms. Come on." As predicted, the lid of the staircase got out of their way as they climbed the flight and permitted them to debouch at the top, but not, as they had expected, on the roof of the single room. They found themselves standing in the middle one of the five rooms which constituted the second floor of the original structure.

For the first time on record Teal had nothing to say. Bailey echoed him, chewing on his cigar. Everything was in perfect order. Before them, through open doorway and translucent partition lay the kitchen, a chef's dream of up-to-the-minute domestic engineering, monel metal, continuous counter space, concealed lighting, functional arrangement. On the left the formal, yet gracious and hospitable dining room awaited guests, its furniture in parade-ground alignment.

Teal knew before he turned his head that the drawing room and lounge would be found in equally substantial and impossible existence.

"Well, I must admit this *is* charming," Mrs. Bailey approved, "and the kitchen is just *too* quaint for words—though I would never have guessed from the exterior that this house had so much room upstairs. Of course *some* changes will have to be made. That secretary now—if we moved it over *here* and put the settle over *there*—"

"Stow it, Matilda," Bailey cut in brusquely. "Wha'd' yuh make of it, Teal?"

"Why, Homer Bailey! The very id—"

"Stow it, I said. Well, Teal?"

The architect shuffled his rambling body. "I'm afraid to say. Let's go on up."

"How?"

"Like this." He touched another button; a mate, in deeper colours, to the fairy bridge that had let them up from below offered them access to the next floor. They climbed it, Mrs. Bailey expostulating in the rear, and found themselves in the master bedroom. Its shades were drawn, as had been those on the level below, but the mellow lighting came on automatically. Teal at once activated the switch which controlled still another flight of stairs, and they hurried up into the top floor study.

"Look, Teal," suggested Bailey when he had caught his breath, "can we get to the roof above this room? Then we could look around."

"Sure, it's an observatory platform." They climbed a fourth flight of stairs, but when the cover at the top lifted to let them reach the level above, they found themselves, not on the roof, but *standing in the ground floor room where they had entered the house.*

Mr. Bailey turned a sickly grey. "Angels in heaven," he cried, "this place is haunted. We're getting out of here." Grabbing his wife he threw open the front door and plunged out.

★

Teal was too much preoccupied to bother with their departure. There was an answer to all this, an answer that he did not believe. But he was forced to break off considering it because of hoarse shouts from somewhere above him. He lowered the staircase and rushed upstairs. Bailey was in the central room leaning over Mrs. Bailey, who had fainted. Teal took in the situation, went to the bar built into the lounge, and poured three fingers of brandy, which he returned with and handed to Bailey. "Here—this'll fix her up."

Bailey drank it.

"That was for Mrs. Bailey," said Teal.

"Don't quibble," snapped Bailey. "Get her another." Teal took the precaution of taking one himself before returning with a dose earmarked for his client's wife. He found her just opening her eyes.

"Here, Mrs. Bailey," he soothed, "this will make you feel better."

"I never touch spirits," she protested, and gulped it.

"Now tell me what happened," suggested Teal. "I thought you two had left."

"But we did—we walked out the front door and found ourselves up here, in the lounge."

"The hell you say! Hm-m-m—wait a minute." Teal went into the lounge. There he found that the big view window at the end of the room was open. He peered cautiously through it. He stared, not out at the California countryside, but into the ground floor room—or a reasonable facsimile thereof. He said nothing, but went back to the stair well which he had left open and looked down it. The ground floor room was still in place. Somehow, it managed to be in two different places at once, on different levels.

He came back into the central room and seated himself opposite Bailey in a deep, low chair, and sighted him past his upthrust bony knees. "Homer," he said impressively, "do you know what has happened?"

"No, I don't—but if I don't find out pretty soon, something is going to happen and pretty drastic, too!"

"Homer, this is a vindication of my theories. This house is a real tesseract."

"What's he talking about, Homer?"

"Wait, Matilda—now Teal, that's ridiculous. You've pulled some hanky-panky here and I won't have it—scaring Mrs. Bailey half to death, and making me nervous. All I want is to get out of here, with no more of your trapdoors and silly practical jokes."

"Speak for yourself, Homer," Mrs. Bailey interrupted, "I was *not* frightened; I was just took all over queer for a moment. It's my heart; all of my people are delicate and highstrung. Now about this tessy thing—explain yourself, Mr. Teal. Speak up."

He told her as well as he could in the face of numerous interruptions the theory back of the house. "Now as I see it, Mrs. Bailey," he concluded, "this house, while perfectly stable in three dimensions, was not stable in four dimensions. I had built a house in the shape of an unfolded tesseract; something happened to it, some jar or side thrust, and it collapsed into its normal shape—it folded up." He snapped his fingers suddenly. "I've got it! The earthquake!"

"Earthquake?"

"Yes, yes, the little shake we had last night. From a four-dimensional standpoint this house was like a plane balanced on edge. One little push and it fell over, collapsed along its natural joints into a stable four-dimensional figure."

"I thought you boasted about how safe this house was."

"It *is* safe—three-dimensionally."

"I don't call a house safe," commented Bailey edgily, "that collapses at the first little temblor."

"But look around you, man!" Teal protested. "Nothing has been disturbed, not a piece of glassware cracked. Rotation through a fourth dimension can't affect a three-dimensional figure any more than you can shake letters off a printed page. If you had been sleeping in here last night, you would never have awakened."

"That's just what I'm afraid of. Incidentally, has your great genius figured out any way for us to get out of this booby trap?"

"Huh? Oh, yes, you and Mrs. Bailey started to leave and landed back up here, didn't you? But I'm sure there is no real difficulty—we came in, we can go out. I'll try it." He was up and hurrying downstairs before he had finished talking. He flung open the front door, stepped through, and found himself staring at his companions, down the length of the second floor lounge. "Well, there does seem to be some slight problem," he admitted blandly. "A mere technicality, though—we can always go out a window." He jerked aside the long drapes that covered the deep French windows set in one side wall of the lounge. He stopped suddenly.

"Hm-m-m," he said, "this is interesting—very."

"What is?" asked Bailey, joining him.

"This." The window stared directly into the dining room, instead of looking outdoors. Bailey stepped back to the corner where the lounge and the dining room joined the central room at ninety degrees.

"But that can't be," he protested, "that window is maybe fifteen, twenty feet from the dining room."

"Not in a tesseract," corrected Teal. "Watch." He opened the window and stepped through, talking back over his shoulder as he did so.

From the point of view of the Baileys he simply disappeared.

But not from his own viewpoint. It took him some seconds to catch his breath. Then he cautiously disentangled himself from the rosebush to which he had become almost irrevocably wedded, making a mental note the while never again to order landscaping which involved plants with thorns, and looked around him.

He was outside the house. The massive bulk of the ground floor room thrust up beside him. Apparently he had fallen off the roof.

He dashed around the corner of the house, flung open the front door and hurried up the stairs. "Homer!" he called out, "Mrs. Bailey! I've found a way out!"

Bailey looked annoyed rather than pleased to see him. "What happened to you?"

"I fell out. I've been outside the house. You can do it just as easily—just step through those French windows. Mind the rosebush, though—we may have to build another stairway."

"How did you get back in?"

"Through the front door."

"Then we shall leave the same way. Come, my dear." Bailey set his hat firmly on his head and marched down the stairs, his wife on his arm.

Teal met them in the lounge. "I could have told you that wouldn't work," he announced. "Now here's what we have to do: As I see it, in a four-dimensional figure a three-dimensional man has two choices every time he crosses a line of juncture, like a wall or a threshold.

Ordinarily he will make a ninety-degree turn through the fourth dimension, only he doesn't feel it with his three dimensions. Look." He stepped through the very window that he had fallen out of a moment before. Stepped through and arrived in the dining room, where he stood, still talking.

"I watched where I was going and arrived where I intended to." He stepped back into the lounge. "The time before I didn't watch and I moved on through normal space and fell out of the house. It must be a matter of subconscious orientation."

"I'd hate to depend on subconscious orientation when I step out for the morning paper."

"You won't have to; it'll become automatic. Now to get out of the house this time— Mrs. Bailey, if you will stand here with your back to the window, and jump backward, I'm pretty sure you will land in the garden."

Mrs. Bailey's face expressed her opinion of Teal and his ideas. "Homer Bailey," she said shrilly, "are you going to stand there and let him suggest such—"

"But Mrs. Bailey," Teal attempted to explain, "we can tie a rope on you and lower you down eas—"

"Forget it, Teal," Bailey cut him off brusquely. "We'll have to find a better way than that. Neither Mrs. Bailey nor I are fitted for jumping."

Teal was temporarily nonplussed; there ensued a short silence. Bailey broke it with, "Did you hear that, Teal?"

"Hear what?"

"Someone talking off in the distance. D'you s'pose there could be someone else in the house, playing tricks on us, maybe?"

"Oh, not a chance. I've got the only key."

"But I'm sure of it," Mrs. Bailey confirmed. "I've heard them ever since we came in. Voices. Homer, I can't stand much more of this. Do something."

"Now, now, Mrs. Bailey," Teal soothed, "don't get upset. There can't be anyone else in the house, but I'll explore and make sure. Homer, you stay here with Mrs. Bailey and keep an eye on the rooms on this floor." He passed from the lounge into the ground floor room and from there to the kitchen and on into the bedroom. This led him back to the lounge by a straight-line route, that is to say, by going straight ahead on the entire trip he returned to the place from which he started.

"Nobody around," he reported. "I opened all of the doors and windows as I went—all except this one." He stepped to the window opposite the one through which he had recently fallen and thrust back the drapes.

He saw a man with his back toward him, four rooms away. Teal snatched open the French window and dived through it, shouting, "There he goes now! Stop thief!"

The figure evidently heard him; it fled precipitately. Teal pursued, his gangling limbs stirred to unanimous activity, through drawing room, kitchen, dining room, lounge—room after room, yet in spite of Teal's best efforts he could not seem to cut down the four-room lead that the interloper had started with.

He saw the pursued jump awkwardly but actively over the low sill of a French window and in so doing knock off his hat. When he came up to the point where his quarry had lost his headgear, he stopped and picked it up, glad of an excuse to stop and catch his breath. He was back in the lounge.

"I guess he got away from me," he admitted. "Anyhow, here's his hat. Maybe we can identify him."

Bailey took the hat, looked at it, then snorted, and slapped it on Teal's head. It fitted perfectly. Teal looked puzzled, took the hat off, and examined it. On the sweat band were the initials "Q. T." It was his own.

Slowly comprehension filtered through Teal's features. He went back to the French window and gazed down the series of rooms through which he had pursued the mysterious stranger. They saw him wave his arms semaphore fashion. "What are you doing?" asked Bailey.

"Come see." The two joined him and followed his stare with their own. Four rooms away they saw the backs of three figures, two male and one female. The taller, thinner of the men was waving his arms in a silly fashion.

Mrs. Bailey screamed and fainted again.

Some minutes later, when Mrs. Bailey had been resuscitated and somewhat composed, Bailey and Teal took stock. "Teal," said Bailey, "I won't waste any time blaming you; recriminations are useless and I'm sure you didn't plan for this to happen, but I suppose you realize we are in a pretty serious predicament. How are we going to get out of here? It looks now as if we would stay until we starve; every room leads into another room."

"Oh, it's not that bad. I got out once, you know."

"Yes, but you can't repeat it—you tried."

"Anyhow we haven't tried all the rooms. There's still the study."

"Oh, yes, the study. We went through there when we first came in, and didn't stop. Is it your idea that we might get out through its windows?"

"Don't get your hopes up. Mathematically, it ought to look into the four side rooms on this floor. Still we never opened the blinds; maybe we ought to look."

" 'Twon't do any harm anyhow. Dear, I think you had best just stay here and rest—"

"Be left alone in this horrible place? I should say not!" Mrs. Bailey was up off the couch where she had been recuperating even as she spoke.

They went upstairs. "This is the inside room, isn't it, Teal?" Bailey inquired as they passed through the master bedroom and climbed on up toward the study. "I mean it was the little cube in your diagram that was in the middle of the big cube, and completely surrounded."

"That's right," agreed Teal. "Well, let's have a look. I figure this window ought to give into the kitchen." He grasped the cords of Venetian blinds and pulled them.

It did not. Waves of vertigo shook them. Involuntarily they fell to the floor and grasped helplessly at the pattern on the rug to keep from falling. "Close it! Close it!" moaned Bailey.

Mastering in part a primitive atavistic fear, Teal worked his way back to the window and managed to release the screen. The window had looked *down* instead of *out*, down from a terrifying height.

Mrs. Bailey had fainted again.

Teal went back after more brandy while Bailey chafed her wrists. When she had recovered, Teal went cautiously to the window and raised the screen a crack. Bracing his knees, he studied the scene. He turned to Bailey. "Come look at this, Homer. See if you recognize it."

"You stay away from there, Homer Bailey!"

"Now, Matilda, I'll be careful." Bailey joined him and peered out.

"See up there? That's the Chrysler Building, sure as shooting. And there's the East River, and Brooklyn." They gazed straight down the sheer face of an enormously tall building. More than a thousand feet away a toy city, very much alive, was spread out before them. "As near as I can figure it out, we are looking down the side of the Empire State Building from a point just above its tower."

"What is it? A mirage?"

"I don't think so—it's too perfect. I think space is folded over through the fourth dimension here and we are looking past the fold."

"You mean we aren't really seeing it?"

"No, we're seeing it all right. I don't know what would happen if we climbed out this window, but I for one don't want to try. But what a view! Oh, boy, what a view! Let's try the other windows."

They approached the next window more cautiously, and it was well that they did, for it was even more disconcerting, more reason-shaking, than the one looking down the gasping height of the skyscraper. It was a simple seascape, open ocean and blue sky—but the ocean was where the sky should have been, and contrariwise. This time they were somewhat braced for it, but they both felt seasickness about to overcome them at the sight of waves rolling overhead; they lowered the blind quickly without giving Mrs. Bailey a chance to be disturbed by it.

Teal looked at the third window. "Game to try it, Homer?"

"Hrrumph—well, we won't be satisfied if we don't. Take it easy." Teal lifted the blind a few inches. He saw nothing, and raised it a little more—still nothing. Slowly he raised it until the window was fully exposed. They gazed out at—nothing.

Nothing, nothing at all. What colour is nothing? Don't be silly! What shape is it? Shape is an attribute of *something*. It had neither depth nor form. It had not even blackness. It was *nothing*.

Bailey chewed at his cigar. "Teal, what do you make of that?"

Teal's insouciance was shaken for the first time. "I don't know, Homer, I don't rightly know—but I think that window ought to be walled up." He stared at the lowered blind for a moment. "I think maybe we looked at a place where space *isn't*. We looked around a fourth-dimensional corner and there wasn't anything there." He rubbed his eyes. "I've got a headache."

They waited for a while before tackling the fourth window. Like an unopened letter, it might *not* contain bad news. The doubt left hope. Finally the suspense stretched too thin and Bailey pulled the cord himself, in the face of his wife's protests.

It was not so bad. A landscape stretched away from them, right side up, and on such a level that the study appeared to be a ground floor room. But it was distinctly unfriendly.

A hot, hot sun beat down from lemon-coloured sky. The flat ground seemed burned a sterile, bleached brown and incapable of supporting life. Life there was, strange stunted trees that lifted knotted, twisted arms to the sky. Little clumps of spiky leaves grew on the outer extremities of these misshapen growths.

"Heavenly day," breathed Bailey, "where is that?"

Teal shook his head, his eyes troubled. "It beats me."

"It doesn't look like anything on Earth. It looks more like another planet—Mars, maybe."

"I wouldn't know. But, do you know, Homer, it might be worse than that, worse than another planet, I mean."

"Huh? What's that you say?"

"It might be clear out of our space entirely. I'm not sure that that is our Sun at all. It seems too bright."

Mrs. Bailey had somewhat timidly joined them and now gazed out at the outré scene. "Homer," she said in a subdued voice, "those hideous trees—they frighten me."

He patted her hand.

Teal fumbled with the window catch.

"What are you doing?" Bailey demanded.

"I thought if I stuck my head out the window I might be able to look around and tell a bit more."

"Well—all right," Bailey grudged, "but be careful."

"I will." He opened the window a crack and sniffed. "The air is all right, at least." He threw it open wide.

His attention was diverted before he could carry out his plan. An uneasy tremor, like the first intimation of nausea, shivered the entire building for a long second, and was gone.

"Earthquake!" They all said it at once. Mrs. Bailey flung her arms around her husband's neck.

Teal gulped and recovered himself, saying:

"It's all right, Mrs. Bailey. This house is perfectly safe. You know you can expect settling tremors after a shock like last night." He had just settled his features into an expression of reassurance when the second shock came. This one was no mild shimmy but the real seasick roll.

In every Californian, native born or grafted, there is a deep-rooted primitive reflex. An earthquake fills him with soul-shaking claustrophobia which impels him blindly to *get outdoors!* Model boy scouts will push aged grandmothers aside to obey it. It is a matter of record that Teal and Bailey landed on top of Mrs. Bailey. Therefore, she must have jumped through the window first. The order of precedence cannot be attributed to chivalry; it must be assumed that she was in readier position to spring.

They pulled themselves together, collected their wits a little, and rubbed sand from their eyes. Their first sensations were relief at feeling the solid sand of the desert land under them. Then Bailey noticed something that brought them to their feet and checked Mrs. Bailey from bursting into the speech that she had ready.

"Where's the house?"

It was gone. There was no sign of it at all. They stood in the centre of flat desolation, the landscape they had seen from the window. But, aside from the tortured, twisted trees there was nothing to be seen but the yellow sky and the luminary overhead, whose furnacelike glare was already almost insufferable.

Bailey looked slowly around, then turned to the architect. "Well, Teal?" His voice was ominous.

Teal shrugged helplessly. "I wish I knew. I wish I could even be sure that we were on Earth."

"Well, we can't stand here. It's sure death if we do. Which direction?"

"Any, I guess. Let's keep a bearing on the Sun."

They had trudged on for an undetermined distance when Mrs. Bailey demanded a rest. They stopped. Teal said in an aside to Bailey, "Any ideas?"

"No... no, none. Say, do you hear anything?"

Teal listened. "Maybe—unless it's my imagination."

"Sounds like an automobile. Say, it *is* an automobile!"

They came to the highway in less than another hundred yards. The automobile, when it arrived, proved to be an elderly, puffing light truck, driven by a rancher. He crunched to a stop at their hail. "We're stranded. Can you help us out?"

"Sure. Pile in."

"Where are you headed?"

"Los Angeles."

"Los Angeles? Say, where is this place?"

"Well, you're right in the middle of the Joshua-Tree National Forest."

The return was as dispiriting as the Retreat from Moscow. Mr. and Mrs. Bailey sat up in front with the driver while Teal bumped along in the body of the truck, and tried to protect his head from the Sun. Bailey subsidized the friendly rancher to detour to the tesseract house, not because they wanted to see it again, but in order to pick up their car.

At last the rancher turned the corner that brought them back to where they had started. But the house was no longer there.

There was not even the ground floor room. It had vanished. The Baileys, interested in spite of themselves, poked around the foundations with Teal.

"Got any answers for this one, Teal?" asked Bailey.

"It must be that on that last shock it simply fell through into another section of space. I can see now that I should have anchored it at the foundations."

"That's not all you should have done."

"Well, I don't see that there is anything to get downhearted about. The house was insured, and we've learned an amazing lot. There are possibilities, man, possibilities! Why, right now I've got a great new revolutionary idea for a house—"

Teal ducked in time. He was always a man of action.

—1964—

SLIPS TAKE OVER

Miriam Allen deFord

Our final story is by Miriam Allen deFord (1888–1975). A prolific writer and lifelong feminist, her reputation as an author rests predominantly on her award-winning mystery fiction and two science fiction collections, Xenogenesis *(1969) and* Elsewhere, Elsewhen, Elsehow *(1971). This story, first published in* The Magazine of Fantasy and Science Fiction, *September 1964, suggests where the mathematical weird would go in the fiction and cinema of the second half of the twentieth century. Strongly suggestive of the physicist Hugh Everett's Many-Worlds Interpretation of quantum mechanics (1957), "Slips Take Over" replaces the notion of extra spatial or temporal dimensions with a conceit for a more assuredly quantum age:* parallel *dimensions.*

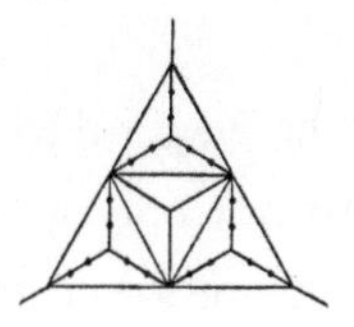

DAVENANT LOOKED UP FROM HIS BEER WITH INTEREST. Words were a hobby of his, and though he had often seen "Bah!" in print this was the first time he had ever heard anybody say it.

"Interviews with Martians—and photographs of them!" the man next to him had said. "Bah!"

Davenant had checked out of his hotel, frugally, to save another day's rent, and had two hours to kill before his plane left for Boston. He was through with the work that had brought him to New York, and he could think of nobody he wanted to call up or go to see. Strolling with his travelling bag in the general direction of the air terminal, he had been brought up short by this little bar he had never noticed before.

"Tim's Place," it said in modest neon; and it had an old-fashioned air. A good place to waste time in, he thought, if it happened to be quiet.

It was quiet enough. There was nobody in the bar at this mid-afternoon hour except the bartender and this bald middle-aged man in a tweed suit. Davenant ordered a beer and had just lifted his glass when he heard that "Bah." He wasn't sure whether the man was talking to the bartender or to him.

It was Davenant he was addressing, His left hand indicating a headline on a newspaper spread before him—which one, Davenant couldn't make out—he gesticulated with a half-full highball glass in his right.

"Science has proved," he went on to the receptive expression on Davenant's face, "that not one of the planets of our solar system is habitable, at least not for creatures like ourselves. The best you could hope for on Mars would be a thinking mushroom. Venus, a thinking fish—a very odd kind of fish. Jupiter, a thinking salamander."

"You don't believe, then, that beings from other planets are watching the earth?" Davenant asked.

"I didn't say that. This universe is full of suns, and a lot of them must have planets revolving around them. Some of those planets may very well be populated by sentient beings. But any civilized—let's say entities—capable of traversing illimitable space would probably be so different from our pattern we wouldn't recognize them as human, or even as individuals. They wouldn't resemble us the least bit, let alone be able to communicate with us.

"No," he went on reflectively, "truth is so much stranger—and so much more familiar. Like this world we're in right now."

"You mean—just our world?"

"I mean *this* world—this frame of reference parallel to the one you come from—I imagine, the same one I did—this one we've both slipped over to."

Davenant gaped at him. The man seemed sober, and perfectly sane.

"I don't get you," he said.

"Look," retorted the bald man. "I can tell. I've never missed a slipover yet. But maybe it's just happened to you and you don't understand.

"You seem to be an educated man. Know any higher mathematics?"

"I ought to. I'm an accountant."

"I don't mean arithmetic. I mean this high-up stuff. Space-time continuum, things like that."

"Sure, I know a little."

"Well, then, didn't you ever hear about multidimensional worlds—parallel frames of reference? I don't know how many there are—nobody does; innumerable ones, possibly. But I do know that in each of them, some few people are so constituted psychologically that the film between is weak—so that they can and sometimes do slip over from one to another. And I miss my guess if you aren't one of them—just as I am."

"I'm afraid you're way beyond my depth," Davenant said.

"No, I'm not. See here." The man in tweeds emptied his glass at a gulp. "Tim!" he called down the bar. "Another of the same for me. And fill up this gentleman's glass."

The burly bartender did the needful, and then stood listening. Davenant nodded his thanks. The bald-headed man went on.

"Ever hear about the farmer who went to his barn to milk his cows, and the cows were found un-milked and the farmer never seen again? Or the private plane that crashed with only the owner in it, and the plane was found, but never the pilot? Or the diplomat who walked around the horses of his carriage—and vanished? Hell, Charles Fort's books are full of cases—supposing you ever heard of Charles Fort. Take Dorothy Arnold, and Judge Crater, and, away back in the early 19th century, Chief Justice Lansing. Where did they all go?

"And how about the people who suddenly turn up on a park bench or on some busy street, years and miles from the life they used to know? Usually they say they can't remember. But where had they been?

"Or take the universal myth, in every country and older than history, of the children stolen by the fairies, or the shepherd who finds a hole in the mountain and enters it. Or Rip Van Winkle. Or the Pied Piper of Hamelin. Myths are just attempts to explain facts without the necessary data.

"And take it another way—what about people like Kaspar Hauser, who suddenly appear—where from?

"So what did happen? In my book, they all slipped over. They slipped into the slot and the zipper closed on them."

"You mean you think they got themselves transplanted into some other dimension?"

"Not the way you probably mean. They couldn't walk out of locked rooms, or turn themselves inside out, or dig holes from the bottom up. But I know darned well they slipped over. I don't know where Crater is—maybe he just got himself murdered—but I've met Miss Arnold—here. She's pretty old by now. And I've seen plenty of others—like you. I recognized the look in your eyes the instant you stepped in here.

"Hell, I ought to know. I'm a slipover myself, as I said. So's Tim here."

Tim nodded solemnly.

Davenant smiled uncertainly.

"Well, it's a good story," he ventured.

The bald man frowned.

"Is that what you think?" he said. "Tell me, a while back, didn't you feel a—a kind of electric shock? In your head? We usually do."

Davenant started. That described it exactly—that funny feeling, just before he'd noticed Tim's Place: like a minor earthquake inside his skull; and then everything seemed to right itself in a second, as if it had been—crooked, before. For a moment he had thought worriedly about high blood pressure, wondered if he'd had a slight stroke. He nodded involuntarily.

"I thought so," said the man.

Davenant got hold of himself with an effort.

"Now wait," he cried suddenly. "I've got you cornered. If this is a different world, how does it happen you speak English?"

"Why not? Don't you? This is New York, isn't it?"

"You mean, you think every city—every place on earth—has its—what did you call it—parallel?"

"Sure. I know they have. I've been in enough of them, in my native world and here."

"So your—your New York has an Empire State Building, and a Rockefeller Center, and a Statue of Liberty, just like mine?"

"I didn't say that. It has the equivalents, but they may not have the same names, or be in the same places, because the history is different. For instance, in our former world I remember there used to be a florist's shop where this bar is now."

Davenant laughed.

"All right, my friend," he said.

"I'll take you right up on that. I'm just down here on business—I live in Boston. Pretty soon I'll be taking a plane home. And I'm willing to bet you anything you like that when I get there Boston will be just where it always has been."

"You can get your plane—though the airport may not be where you expect. And you'll reach Boston on schedule. Boston Harbor will be there, and Beacon Hill, and the Charles River—all natural objects. But they might not be called by the same names (I don't know—I've never been in Boston in this world), and all the buildings will be different. And in the whole city there won't be one human being you ever saw before—unless you meet another slipover."

"And even if he did meet another," the bartender interpolated, "he might not be from the same world originally, Mr. Gorham. You and me aren't."

"That's right. I didn't think of that. I've got a hunch, though, that you and I did come from the same place, Mr.—"

"Davenant. Charles Davenant."

"My name's Gorham—James B. Tell me, Mr. Davenant, did you ever hear of Aristotle, or Julius Caesar, or William the Conqueror, or Shakespeare?"

"Are you kidding?"

"O.K. Tim—you ever hear any of these names before?"

"Now, Mr. Gorham, you know I ain't had much education."

"All right, then tell me—who was Lincoln? Who was Washington? Ever hear of Hitler? Or Stalin? Or Eisenhower?"

"You've got me," Tim said soberly.

"You see? You and I have the same history—Tim hasn't. The great names he knows we wouldn't recognize. But he came from *his* America, just as we came from ours."

"But once in a while I do meet somebody from my own place," Tim put in eagerly, and then we remember the same things. Like, Randolph took Richmond in the Civil War, or Thomas Endicott was the first president… It's never anybody I ever knew before, though."

"See?" said Gorham. "That's the way it goes, Mr. Davenant. History gets changed a little in each world.

"Few slipovers, relatively—in numbers, many. Hell, people disappear from every big city every day. If they happen not to have friends or relatives to notice or care about them, they're never even missed. You married?"

"No," said Davenant uncommunicatively. He was thinking.

"That's good. The worst part of it all, the way I see it, is the wife or husband left to wait and wonder and never know what happened. It's worse for them than for the one that slips over, for at least he knows he isn't dead and didn't desert. I was lucky that way too—though I'd give anything to be able to let my mother and dad know I didn't just run out on them.

"It's funny—sometimes more than one member of a family is the special type that can slip over. I heard of two brothers, out in Oakland, California. They both slipped over, four years apart; both did it the same way—walked out of their house—two old bachelors they were—leaving the lights burning, the radio going, dinner on the stove. When the second one arrived, they found each other. If they're not dead, they're still together in Oakland—*this* Oakland.

"But I've never heard of a married couple who were both slip-overs. They say opposites attract—perhaps slipover types never marry each other. Sometimes when a man or woman has been here a long time and seems likely to stay here, he or she remarries. It's bigamy, of course—but the law will never catch up with them. I'm married now myself—but then I never was before."

Davenant stared at the two men.

"You really believe all this stuff?" he asked slowly.

Gorham sighed.

"I know—it took me a long, hard time too. That's why I try now to help others, when I recognize them.

"Haven't you noticed that nobody's walked in here since you did? It's not that quiet, even at this hour, eh, Tim? I didn't want us to be interrupted. I gave Tim the wink while you had your back turned and he locked the door so we could have a long talk. This is his own place—he's boss."

"That's right," said Tim. "Mr. Gorham's been a good friend to me—helped me buy this joint. I don't mind losing a little trade once in a while to do him a favour."

Davenant felt the blood rushing to his head.

"Hey!" he yelled. "I don't like this! Let me out, or—"

"Easy does it, fellow. You can walk out any time you want. We won't stop you.

"But look, let's discuss this quietly a little, shall we? Have another beer, and go ahead and ask me any questions you want."

Davenant's momentary anger left him. He could be a good sport and go along with a joke. He glanced at his watch. Plenty of time yet.

"O.K.," he said. "What about clothes? Or this bag of mine?"

"Your clothes were on you, and they came over with you. It isn't like teleportation. But look and see if you've got a return ticket to Boston. You won't have, because you didn't buy one here."

Davenant pulled his fingers away from his empty pocket as if they had been bitten.

"It's some sort of sleight-of-hand," he muttered. "I can feel the money still in my wallet."

"Why not? You had that on you too—though you can't spend it here. You can exchange it for as much as you need of mine. It will look different, but it will be good, and I can keep yours as a souvenir."

Interesting new con game, Davenant thought. Gorham seemed to read his mind.

"Listen, Mr. Davenant, if you think I'm playing a silly joke on you, I can prove to you who I am."

He began producing identification—driver's licence, Chamber of Commerce membership, credit cards.

"I want to help you, my friend. Nobody helped me, at first, and I know how tough it is. Say you go to Boston, and for the sake of the argument say you find things the way I've told you. You won't have your home or your job—they're off somewhere in another parallel frame of reference. See here—"

He held out a business card. James B. Gorham, assistant vice-president, Bank Mutual Life Insurance Company.

"We can use another accountant in our Boston office. You'd have to qualify, of course. But you can refer to me, and that will get you over the worst hurdle for every slipover—not having any proof of degrees or experience."

Davenant looked suspiciously at the card in his hand.

"Never heard of the company," he remarked.

"It's an old-line one," said Gorham equably. He pointed to a printed statement: "Established 1848."

Something occurred to Davenant. His face brightened with triumph.

"Got you at last!" he chuckled. "So you're a 'slipover' yourself, are you? *You* didn't have any credentials, either, when you came. So how come all at once you're assistant vice-president of a big insurance company?"

"Not all at once, Mr. Davenant." Gorham's voice was dreary. "I've spent half a lifetime here by now. I guess I'll die here. I don't know that I'd even want to go back any more—I've forgotten a lot, and most of the people I knew there would be dead."

"Well, what about the people who do go back?" Davenant demanded. "Why don't they tell what happened to them? Why do they always have amnesia, when *you* don't have it here for the—for your other world?"

"Why do they always *say* they can't remember, you mean. Maybe a few really do have amnesia from shock. But just consider a minute. What would happen to anybody who'd slipped over and back, and then tried to explain the truth? For that matter, what would happen to anyone, in this or any other parallel world, who would tell the truth to anybody except another slipover? How long would it take to put the raving maniac in a mental hospital? I guess plenty of them are there right now as it is, poor devils.

"And think how much worse it would have been in the days before people had any scientific concepts. Think of the fate of any poor fool then who told where he'd been or where he'd come from: chained to an iron bar on a heap of straw, or burnt at the stake as a witch."

"Now, wait a minute," Davenant objected. "You're saying the civilizations are the same in all these so-called parallel worlds of yours? You mean—say this really is a different world we're in, it's in the same fix today as ours—threatened with nuclear war and destruction and chaos and all the rest of it?"

"I said they were parallel worlds, my friend," Gorham replied gravely. "The history is different in detail, but in the end, like causes lead to like effects. As for your second question, unfortunately the answer is yes, at least so far as this world is concerned.

"But you've got a problem of your own to solve before you have time to discuss politics or sociology. Ask me anything you want to about that. And any time you say, Tim will open the door and you can walk out and try to find your way to Boston."

"If this whole thing is an elaborate practical joke," said Davenant painfully, "I give in; you've made a fool of me and let's call it quits.

"But all right; I'll play along with you some more. Why do some of these people who disappear turn up again right away, in a few hours maybe, while others never come back at all?"

"I don't know why; I just know they do. Some people slip back and forth frequently, and learn to manage it. And I've met a few who made such a short transition—'translation into the positive absolute,' Fort calls it, whatever that means—that they hardly realized it themselves. Maybe you'll be one of the short-timers; I hope so, for your sake.

"I've never heard of anybody who slipped over into more than *one* other world, but perhaps there are some of those too. For the quick back-and-forthers, it may seem that they've just had a vivid dream, if it happened while they were asleep. If they were awake, the double shock might be too great and they might just blank out and forget the whole thing. Or it might even kill them. Perhaps that's what happens to some of the people who are found dead in bed, with no evidence of disease."

Scared and sick, Davenant stared at Gorham. He was remembering things. He groped his way to a chair by one of the little tables and sat down.

One night, when he was a very small boy, he had had a strange dream that he could still remember. In his dream, he was walking down a street, when suddenly he heard a dull rhythmic booming. He asked a woman passing by what that was, and she answered, "That's the washerwomen who live down under the earth."

A child's rationalization. But he was in that period of life when illusion and reality are inextricably mixed. So soon afterwards he asked his mother, "Why don't I hear the washerwomen any more?" "What are you talking about?" she asked, and he explained. She laughed. "You just dreamed that, dear," she said... But he never forgot.

He remembered something else. Often, as he grew older, he had a strange experience just before he fell asleep. Unknown faces would suddenly flash before his consciousness, or he would catch scraps of conversation that he could never recall. Before he was grown, he had in his own mind divided both sight and hearing into three categories—ordinary vision and sound, purely imagined or remembered vision and sound, and what he called "the inbetween." He supposed everybody shared his experience, till one day he mentioned these hypnagogic experiences casually to his chum. "Are you crazy, Chuck?" Russell wanted to know. "Why," he answered, astonished, "don't you have it too?" "Have what? You cuckoo or something?" After that, he never spoke of "the inbetween" again.

Gorham and Tim were watching him compassionately. He rose shakily to his feet. "So that was what—" he began quaveringly.

And then all at once he recalled something. He felt his face turning white. He had never been so angry in all his life.

"Interviews, and photographs of Martians!" he choked. "Of all the dirty tricks!" How grown men could get a kick out of playing a rotten joke like that—trying to kid me into believing—

"Listen, you! It's only a few years since those flying saucer books began to come out. You've been 'here' for years and years, have you? Then where did you learn about those nuts who think they've met extraterrestrials? And don't tell me you've been conversing with some other 'slipover' that just got here and told you about it. That's not the sort of thing that would be likely to come up in an ordinary conversation!"

"I said these were parallel worlds, Davenant," said Gorham quietly. "They have parallel myths, too."

"Rats! Let me out of here! Now—this minute."

"Certainly. Let him out, Tim."

Tim walked around the bar, reached into his trousers pocket, and imperturbably unlocked the door. Then he blocked the way and held out a broad palm.

"That'll be 50 cents, mister, for that first beer," he announced.

Scarlet with embarrassment Davenant pulled a bill from his wallet, noted the "1" on the corner.

"Keep the change for the floor show," he growled.

For a second, then, he almost snatched the money back. "Nuts!" he mumbled. He pulled the door open and slammed it behind him, too furious to glance back at the pitying faces, too furious to do anything but march rapidly down the street toward where, he knew, the west side air terminal was.

It was right there. Did the airport bus look a little different? Everything was going to look a little different now; Gorham had thoroughly upset him with his nonsense. But the plane looked just like the one that had brought him down here, and so did the bus from the Boston airport to the city.

He needn't go to the office till tomorrow; he'd phone from his his bachelor apartment on the wrong side of Beacon Hill. He hailed a taxi, and noted with an unpleasant shock that it was pink. Had he ever seen any pink taxis before? Well, he wasn't very observant, and there were always new taxi companies starting up. They'd gone almost the whole distance before he realized he was keeping his eyes away from the window. At the same moment the driver spoke.

"Did you say Number 12, Mister? There ain't no Number 12 on Laurel Street."

It was Laurel, all right; he recognized some of the houses. But where his apartment house had stood there was a parking lot.

Davenant felt a little sick. He'd sort this all out soon, but now he had to get somewhere where he could be alone and sit down and think things over. "Take me to the Copley-Plaza," he said in a strangled voice.

"Mottley-Plaza it is," said the driver. Davenant shuddered.

He wouldn't look; he wouldn't notice the differences. He got a room without difficulty and followed the bellboy numbly into the elevator and down the hall.

"Hey!" said the boy, about to leave, "what kind of funny money is this?"

Davenant dared not glance at the half dollar he had just given the boy. Whose head should be on it instead of Franklin's? He tried to smile but the smile turned into a grimace. The boy looked a little frightened. "Cheap skate!" he muttered under his breath, and left quickly. Davenant locked the door.

"Get hold of yourself!" he admonished himself sternly. He took off his tie and doused his head in cold water. When he stopped shaking he set his jaw and lifted the phone. He gave the operator the familiar number of his office.

He hung up, and the phone rang almost immediately. With his heart beating too fast, he said: "This is Davenant. Put George Watson on, Lucille." A voice broke in; it was the hotel switchboard girl again.

"I'm sorry, sir, but I got a recording that the number you called is not a working number."

Suddenly he was very angry.

"Look," he snapped, "I'm calling Black, Watson, and Heilkrammer, in the Old State Building. Maybe they've changed their number overnight, but I don't think so. Get them for me."

"I'm sorry, sir, it isn't my—" But he had hung up again. This time it was nearly five minutes before she called him back.

"There is no Black, Watson, and Heilkrammer listed in the phone book. And there is no Old State Building in Boston."

Davenant cradled the phone without another word. He sat back in his chair, his head whirling.

Even supposing that preposterous nonsense of Gorham's had been the truth, then how had he been able to get here at all? Why had the man in the air terminal in New York taken his money for the ticket? Oh-oh—now he remembered. He had cashed a traveller's cheque; presumably they were the same in both worlds. And the taxi driver—he must have paid him from the change he got at the terminal. But the bellboy's tip had come from another pocket; it was money he had had on him before he— Before.

Wait: there was one way to get the thing straightened out, or as much as it could be straightened out for the present. He fished in his wallet for the card Gorham had given him. Bank Mutual Life Insurance Company, James B. Gorham, assistant vice-president. He read it aloud to see if he could talk without his voice trembling; then, his lips and fingertips cold, he lifted the phone again and gave the number on the card.

He was not surprised—only scared to his very depths. Somehow he had almost expected it.

"That isn't a working number either, sir." The switchboard operator hesitated. "Excuse me, but these *are* Boston numbers you're calling?"

"Never mind," he managed to breathe, and he got the receiver back in its cradle. Something had just occurred to him.

He recalled his angry exit from Tim's Place, he recalled walking indignantly away and down the street. And now something else came back to him. Somewhere between the bar and the terminal, that strange thing had happened to him again: that tiny instantaneous explosion, like a small electric shock, piercing his brain; then suddenly things seemed to right themselves again.

But where? Into what world had he slipped then? Where in God's name was he now?

He turned his face to the back of the armchair and clung to its sides. Dry sobs shook him and his throat felt raw.

"Help me!" cried Davenant to somebody or something, a lost child. "Help me! I want to go home!"

British Library Tales of the Weird collects a thrilling array of uncanny storytelling, from the realms of gothic, supernatural and horror fiction. With stories ranging from the nineteenth century to the present day, this series revives long-lost material from the Library's vaults to thrill again alongside beloved classics of the weird fiction genre.

EDITED BY MIKE ASHLEY

Doorway to Dilemma
Bewildering Tales of Dark Fantasy

From the Depths
And Other Strange Tales of the Sea

Glimpses of the Unknown
Lost Ghost Stories

The Outcast
And Other Dark Tales by E. F. Benson

A Phantom Lover
And Other Dark Tales by Vernon Lee

The Platform Edge
Uncanny Tales of the Railways

Queens of the Abyss
Lost Stories from the Women of the Weird

EDITED BY DAISY BUTCHER

Evil Roots
Killer Tales of the Botanical Gothic

EDITED BY GREG BUZWELL

The Face in the Glass
The Gothic Tales of
Mary Elizabeth Braddon

Mortal Echoes
Encounters with the End

EDITED BY ELIZABETH DEARNLEY

Into the London Fog
Eerie Tales from the Weird City

EDITED BY TANYA KIRK

Spirits of the Season
Christmas Hauntings

Chill Tidings
Dark Tales of the Christmas Season

EDITED BY JOHN MILLER

Tales of the Tattooed
An Anthology of Ink

Weird Woods
Tales from the Haunted Forests of Britain

EDITED BY XAVIER ALDANA REYES

Promethean Horrors
Classic Tales of Mad Science

Roarings From Further Out
Four Weird Novellas by
Algernon Blackwood

The Weird Tales of
William Hope Hodgson

EDITED BY ANDREW SMITH

Haunted Houses
Two Novels by Charlotte Riddell

We welcome any suggestions, corrections or feedback you may have, and will aim to respond to all items addressed to the following:

The Editor (Tales of the Weird), British Library Publishing,
The British Library, 96 Euston Road, London NW1 2DB

We also welcome enquiries through our Twitter account, @BL_Publishing.